Air and Space Power

Significant Issues Series

SIGNIFICANT ISSUES SERIES papers are written for and published by the Center for Strategic and International Studies.

Director of Studies: Erik R. Peterson

Director of Publications: James R. Dunton

Managing Editor: Roberta L. Howard

The Center for Strategic and International Studies (CSIS), established in 1962, is a private, tax-exempt institution focusing on international public policy issues. Its research is nonpartisan and nonproprietary.

CSIS is dedicated to policy analysis and impact. It seeks to inform and shape selected policy decisions in government and the private sector to meet the increasingly complex and difficult global challenges that leaders will confront in the next century. It achieves this mission in three ways: by generating strategic analysis that is anticipatory and interdisciplinary; by convening policymakers and other influential parties to assess key issues; and by building structures for policy action.

CSIS does not take specific public policy positions. Accordingly, all views, positions, and conclusions expressed in this publication should be understood to be solely those of the individual authors.

❖ ❖ ❖

The Center for Strategic and International Studies
1800 K Street, N.W.
Washington, D.C. 20006
Telephone: (202) 887-0200
Fax: (202) 775-3199
E-mail: info@csis.org
Web site: http://www.csis.org/

Air and Space Power in the New Millennium

Edited by *Daniel Gouré and Christopher M. Szara*
Introduction by *Gen. Ronald R. Fogleman, USAF*

Published in cooperation with VII Inc.

THE CENTER FOR STRATEGIC & INTERNATIONAL STUDIES
Washington, D.C.

Photo credits—Front cover (left to right): F-22 Raptor, courtesy of Lockheed Martin Corporation; F/A-18E/F Super Hornet, courtesy of the Boeing Company; V-22 Osprey, courtesy of the Department of the Navy, U.S. Marine Corps, Public Affairs Office; RAH-66 Commanche, courtesy of Boeing/Sikorsky Commanche Project.

Back cover (left to right): E-8 JSTARS, courtesy of Northrup-Grumman Corporation; ATACMS Missile, courtesy of the Department of the Army, Public Affairs Office; Predator UAV, courtesy of General Atomics Aeronautical Systems, Inc.; AH-64D Longbow Apache, courtesy of the Boeing Company.

Cover design by Meadows Design Office.

Significant Issues Series, Volume XIX, Number 4
© 1997 by The Center for Strategic and International Studies
Washington, D.C. 20006
All rights reserved
Printed on recycled paper in the United States of America

99   98   97   4   3   2   1

ISSN 0736-7136
ISBN 0-89206-330-0

Library of Congress Cataloging-in-Publication Data

Air and space power in the new millenium   / edited
   by Daniel Gouré and Christopher M. Szara.
         p.   cm. — (Significant issues series, ISSN 0736-7136 ; v. 19, no. 4)
   Includes bibliographical references.
   ISBN 0-89206-330-0
       1. Air power.   2. Astronautics and civilization.
   3. Twenty-first century.   4. Air power—United States.
   I. Gouré, Daniel.   II. Szara, Christopher M.   III. Series.
UG630.M374   1997
358.4--dc21   97-29795
                                                                                   CIP

Contents

Foreword *Saxby Chambliss, Member of Congress*   vii

Acknowledgments   *ix*

Executive Summary   *xiii*

Steering Committee   *xxv*

Introduction *Gen. Ronald R. Fogleman, USAF (Ret.)*   *xxvii*

1
The Coming of Age of Air and Space Power   *1*
    Daniel Gouré and Stephen A. Cambone

2
Strategy   *48*
    Jeffrey R. Cooper

3
Air and Space Superiority   *88*
    Richard P. Hallion and Michael Irish

4
Global Attack and Precision Strike   *106*
    Jeffry A. Jackson

5
Information Superiority   *117*
    C. Edward Peartree, C. Kenneth Allard, and Carl O'Berry

vi *Contents*

6
Mobility and Support   *132*
   Keith A. Hutcheson and Robert McClure

7
Technology and the Industrial Base   *143*
   Ivars Gutmanis

Abbreviations and Acronyms   *159*

Working Groups   *163*

About the Contributors   *167*

Foreword

When we look at the approaching millennium through the lens of national security, we can be certain of at least one thing: the strategic environment will look nothing like what we have experienced in the past or have even planned for. Traditional conflict fought on the battlefields of Verdun, Iwo Jima, and most recently the desert of Southwest Asia will be supplanted by warfare fought from afar, most notably from space, by electronic means and with standoff weaponry. Moreover, the technological dominance we have enjoyed over our adversaries will be offset by greater access to weapons of mass destruction and formidable skills in conducting information warfare. Indeed, our near-term and midterm foes will not likely be peer competitors but will be smaller state and nonstate actors that can threaten U.S. interests at home and abroad with mobile, dispersed, and decentralized forces. If one accepts this notion of a changing and unpredictable strategic environment, then we must ask how do we prepare for such conditions? What forces are necessary to meet these challenges? The most compelling argument resides in this volume.

In *Air and Space Power in the New Millennium* warfare is placed in an entirely new paradigm. When we send our sons and daughters to war, we expect our military forces to prevail decisively and with as few casualties as possible. While the skills of U.S. ground and naval forces fighting in the horizontal would likely conquer any adversary, this type of warfare is costly in time and in lives. The two-dimensional—horizontal—battlefield has its place in history; the focus has now shifted into a third dimension—the vertical. This is the realm of air and space forces.

When we control the third dimension we control both the horizontal and the vertical battlefields. Bases of operations placed in the third dimension afford the enemy nowhere to hide and allow us to "see" the entire battlefield from a position that ensures a successful campaign. This equates to a contained conflict, fought on our terms, of limited duration, and with fewer

American sailors, soldiers, airmen, and marines exposed to lethal fire.

During the last half century, the American people paid the price for this nation to win the Cold War and become the sole remaining superpower. As a consequence, expectations run high; the bar has been raised. We often speak in terms of air superiority. In future conflict, superiority is not enough: air and space dominance must be our objective. As cochair of the Congressional Air Power Caucus, I commend this volume to this nation's military decision-makers so that they may better understand the ever-increasing role of air power in coming conflict.

THE HONORABLE SAXBY CHAMBLISS
United States House of Representatives

Acknowledgments

A study of such ambitious scope as *Air and Space Power in the New Millennium* could not have been pursued successfully without the participation, support, assistance, and guidance of a great number of people. More than 150 individuals attended the study's meetings and conferences as either participants or speakers. During the course of the study, many of them provided written comments, briefings, or data to support the study process. Many others worked behind the scenes to support the effort.

The study members would like to acknowledge the invaluable role played by the members of the Steering Committee. They provided the critical guidance, review, and commentary that ensured that the final product would be a worthwhile contribution to the national debate on U.S. military strategy for the twenty-first century. Many steering committee members took the time to provide valuable oral and written critiques of the study that helped focus the effort and refine the arguments.

We would like to thank Maj. Gen. Charles Link, USAF (Ret.) and Lt. Col. Michael Murphy, USAF, for their support in allowing CSIS to explore this issue in an open and joint environment. Special thanks are extended to the following individuals for their extra effort and intellectual contribution to this study: Adm. James Hogg, USN (Ret.); Mr. Frank Jenkins; Gen. Carl Mundy, USMC (Ret.); Mr. Robert Murray; Lt. Gen. Bernard Trainor, USMC (Ret.); and Dr. Harlan Ullman. We would also like to thank those who contributed briefings, papers, and lectures to this study's many working group meetings.

Each of the six working groups held an average of six formal meetings over a period of four months. Members of the working groups gave generously of their time and talent in preparing the series of reports that formed the intellectual foundation of the study. Our thanks go out to every participant of the working groups. We are particularly grateful to those who agreed to chair the individual working groups: Strategy, Gen. Michael Dugan, USAF (Ret.) and Adm. Ronald "Zap" Zlatoper, USN (Ret.); Information Superiority, Vice Adm. Michael McConnell, USN (Ret.); Global Attack and Precision Strike, Mr. Arnold Punaro; Air and Space Superiority, Dr. Richard Hallion; Mobility and Support, Gen. H. T. Johnson, USAF (Ret.) and Brig. Gen. Tom Mikolajcik, USAF (Ret.); and Technology and Industrial Base, Mr. Hal Howes. Their leadership and wisdom were critical to the success of the study and the quality of the working group products.

Selected members of each working group distilled from the working-group sessions the essence of the arguments on the particular merits of air and space power: Strategy, Mr. Jeffrey Cooper; Information Superiority, Mr. Edward Peartree, Dr. Kenneth Allard, and Lt. Gen. Carl O'Berry, USAF (Ret.); Global Attack and Precision Strike, Lt. Col. Jeffry Jackson, USAF; Air and Space Superiority, Dr. Richard Hallion and Mr. Michael Irish; Mobility and Support, Mr. Keith Hutcheson and Lt. Col. Robert McClure, USA; and Technology and Industrial Base, Dr. Ivars Gutmanis.

The study owes a special debt of gratitude to the CSIS staff, particularly the members of the Political-Military Studies Program. This study would not have been possible without their tireless efforts. Christopher Szara, coeditor of this volume, and Jennifer Metzler undertook the herculean task of organizing and coordinating study activities. Each of our military fellows, who come to CSIS for a one-year tour of duty, participated in the study, often devoting days and weeks to the effort. We are grateful to each of them: Lt. Col. Dewey Mauldin, USMC; Lt. Col. Jeffry Jackson, USAF; Lt. Col. Robert McClure, USA; and Comdr. Richard Dawe, USN. Behind the scenes, the research assistants and interns in Political-Military Studies gave unstintingly of their time and energies. We are especially grateful to Joe Cyrulik, John Kreul, Alisa Mandel, John Swann, and Lesley Young. Our conference department, Janet Granger and Carla Corliss, shouldered the burden of supporting some 40 meetings.

Finally, we wish to recognize the role of one individual without whom this effort would not have been possible. Gen. Ronald Fogleman, USAF (Ret.), former chief of staff of the United States Air Force, had the wisdom to see the need for a debate on U.S. military strategy for the next century, the determination to promote that debate, and the courage to see it conducted in a fair and unbiased manner. It is to be hoped that by his efforts he has set the standard for what will inevitably be an extended discourse among the American people regarding future national security policy and the manner in which they wish to see the interests of this nation protected.

DANIEL GOURÉ
Study Director, CSIS

CHARLES D. VOLLMER
President, VII Inc.

Executive Summary

What is modern air and space power? Modern air and space power is the ability to conduct military operations simultaneously and globally in three dimensions from a base of operations in the third dimension—air and space. In strategic terms, it requires the ability to control and exploit the air and space medium. This, in turn, requires the ability to deploy, operate, command, and control effective offensive, defensive, and movement capabilities in and through the air and space mediums while denying, as necessary, those capabilities to an adversary. Modern air and space power consists of space systems, including the means necessary to place objects in space; air systems, both manned and unmanned fixed-wing and rotary platforms; missile systems operating from above, on, or—in the case of underwater platforms—below the surface of the Earth against targets in air or space or on the surface; and the command, control, communications, computers, intelligence, surveillance, and reconnaissance (C^4ISR) systems that enable linkage to all of these systems.

Laymen and professionals alike too often associate U.S. air power solely with fixed-wing aircraft, principally those operated by air forces. While these are critical elements of this nation's air capability, they are not the only elements that contribute to the creation of modern air power. United States Navy carrier battle groups, surface ships, and submarines armed with cruise missiles; United States Army artillery battalions armed with ATACMS (Army Tactical Missile System) ballistic missiles; United States Marine Corps and special operations forces V-22s and the helicopters operated by each of the services—all are examples of elements of the air power equation. Tying all these systems together is a vast array of intelligence collection, communications, and support systems. When it comes to space power, each of the services has a role. In some areas, such as satellite-based communications, modern air and space power is based increasingly on the use of commercial assets.

Why a study of the merits of modern air and space power? There is no question that air and space power constitutes one of the strongest elements of U.S. military power: it is a demonstrably unique U.S. national competence that provides an asymmetric advantage over potential adversaries. The role of air and space power in diplomacy, crisis, and war has changed and expanded dramatically over the past century, as air and space forces have developed the capability to perform more missions, reach out farther, and strike with more force. It is the character of that power and the extent to which it can be made central to strategic and operational planning that have provoked a strong debate. This debate raises a number of important questions. Does air power have an independent role? Is it now time for the services, including the United States Air Force, to transform their doctrine and force structure to a new concept of warfare? Can we transform warfare, lessen the bloodiness of "direct contact," and enhance our capacity to prevent or deter hostile actions through the exploitation of air and space forces? We have been debating these questions for nearly 50 years. Until now the answer has been no.

This study was initiated to assess whether that answer should be reconsidered and to inquire whether the claims for air power now rest on a sound footing. If they do, air and space power, given its potential, ought to be treated not as another element of the restructured force but as its fundamental building block, the keystone, allowing the creation of armed forces capable of fighting and winning the nation's wars. Although we are in a time of shifting national security priorities, this study was established with the belief that the central measure of merit for U.S. military forces as a whole must be their ability to prevail in wartime. For air and space power, the measure of merit must be the opportunities it provides to ensure that the nation, its interests, and its citizens can be defended efficiently and effectively.

In lesser contingencies the role of air and space power will be determined by the specifics of the situation. Air and space power was not the dominant player in Somalia or Haiti. Nevertheless, it played a critical role in supporting forces deployed to those two nations.

Overall, however, in an era of changing national security concerns, when fewer resources are devoted to defense, the contributions of air and space power within the restructured armed forces may now be such as to permit air- and space-related forces to play new and critical roles, not only in meeting current security requirements but also in helping shape a security environment for the decades ahead.

Since the end of the Cold War, America's air and space forces have been reduced substantially, as indeed have all elements of the armed forces. If modern air and space power can constitute the fundamental building block for the military of the twenty-first century, it is time to consider how the strength of remaining air and space forces can be improved and their capabilities more fully exploited to create a joint war-winning capability.

Confronting the Challenges of a New Millennium

The adequacy of U.S. forces to support national security and military strategies must be evaluated relative to a range of challenges to U.S. security and global interests that are likely to be more difficult than those seen in the near-decade since the Cold War ended. In addition, the context of U.S. involvement is likely to be different from that of the Cold War, when U.S. forces were forward deployed defending well-established borders. Finally, potential U.S. adversaries clearly are not standing still. They are taking advantage of opportunities presented in the international arms market.

It seems increasingly likely that the future international security environment confronting the United States will be more complex in a number of dimensions. Whereas current scenarios envision adversaries to be rather limited in size, capabilities, and reach, this may not be the case in the future. The Quadrennial Defense Review (QDR) proposed a future security environment that suggested the need to prepare to meet the threat of robust regional adversaries early in the next century and the prospect of heavily armed theater-level "peer" competitors or major powers by approximately 2014. Although these major powers are unlikely to have the means to pose a global challenge to the United States, they will have sufficient capability to compete with the United States and its allies within their own theaters of regard or influence. In that context they will constitute theater-level peers that potentially approach a "continental-size adversary."[1] In World War II, neither Germany nor Japan was a global power, and in the end neither had sufficient power to win against the weight of the combined forces of the Allies. Yet both conducted continental-size wars for a period of years. Theater-level peers are likely to have sufficient resources and occupy enough territory to become resilient adversaries. In addition, they are likely to possess weapons of mass destruction (WMD) and the long-range delivery systems necessary to threaten distant U.S. interests and even the U.S. homeland. They may be members of an alliance structure, creating the possibility that they too can mobilize coalitions in opposition to a coalition led by the United States.

Potential adversaries, whatever their size, will likely be tougher politically, operationally, and technologically in the future than they have been in the past. They are likely to possess WMD and delivery systems in militarily significant numbers and configurations. They will employ their forces jointly and have access to C^{4}ISR capabilities better than much of what the United States deployed only a few years ago. They are currently attempting, some more successfully than others, to acquire advanced military capabilities that could pose a challenge to the United States. Among the advanced capabilities that will be of greatest concern for U.S. in-theater forces are airborne and space-based surveillance, communications, and navigation and targeting systems coupled to long-range delivery systems (such as tactical ballistic missiles and cruise missiles). Many of these capabilities are or soon will be available in the commercial or military markets.

The presence of so-called new threats does not obviate the need to be prepared to deter and, if necessary, fight and defeat rogue states, hostile regional powers, and potential peer competitors. What is important about modern air and space power is that it is increasingly vital to the nation's ability to address both the broadest portion of that threat spectrum and do so in a manner that can increase expectations that the United States will be able to meets its security and defense objectives with the least risk. Air and space power appears to offer increased response options to decision-makers and military leaders confronted by an uncertain threat, changing political and military conditions in zones of conflict, the need to wait until the adversary makes the first move, and the requirement to project power rapidly over long distances into theaters of widely varying character. It should be noted, however, that the ability of modern air and space systems to meet the challenges posed by military threats of widely differing character and intensity suggests that some of those same capabilities—or new systems derived from them—

should be capable of addressing requirements posed by many of the so-called new threats.

The Revolution in Air and Space Power

The Persian Gulf War, with its demonstration of the power of new technologies (stealth, precision guidance, airborne surveillance, space-based sensors and communications) and of the capability of modern battle management and command, control, and communications (C^3) systems to control and direct them, sparked a debate about to whether air power had reached maturity and whether the "coming of age of air and space power"[2] meant that the independent air operation was a realistic prospect and could have strategic effects. The ability to control the skies and to deliver precision ordnance with impunity virtually anywhere in the theater marked a qualitative leap in the capability of modern air power. Although precision weapons constituted only 9 percent of the total tonnage dropped during the Gulf War, they accounted for more than 42 percent of the damage to strategic targets.[3] A handful of precision-capable aircraft also inflicted a disproportionately large number of casualties on Iraqi field forces.

The complete realization of air and space power's strategic potential is possible today because, for the first time, air and space power can deliver almost fully on the vision put forward by the early air power advocates. In the past, technical limitations in target identification, navigation, and weapons delivery often meant that even massive air operations produced little in the way of tangible results. These limitations often resulted in a need to mass platforms over the target. The relative strength of anti-air defenses, the inherent vulnerability of air platforms, and the requirement to fight to and from the target when air superiority was not secured often created extremely high losses in strategic bombing missions. The synergy of improved intelligence and surveillance, accurate navigation, precision targeting and weapons delivery, and reliable real-time communications greatly improved survivability; and the availability of a variety of delivery means has fundamentally changed the character of warfare in and from air and space.

What is truly revolutionary about modern air and space power is its potential to fundamentally change both the way we will perceive and the way we will conduct warfare in the future.

For millennia, mankind has lived, fought, and died on the surface of the Earth. For many decades, even man's ability to move through the air and under the seas did not change his two-dimensional perception of his existence. Today, this perception is changing at an ever-increasing rate as technology allows us to exploit the opportunities created by operating in and through air and space. Our historical, horizontal perspective is giving way slowly to a new, vertical perspective. As we grow comfortable with this change, we grow increasingly comfortable as well with seeing our reality in three dimensions instead of two.

The same revolution of perspective and viewpoint has influenced warfare. As the Gulf War suggested and has been clarified in the ensuing years, technology is liberating warfare from the horizontal perspective. Experience in the Gulf War suggested the potential for achieving a new perspective. For example, the effort to counter Iraqi Scud missiles was conducted from a vertical perspective. What this experience revealed was that the United States has the ability to perceive the world from a vertical perspective and to communicate the knowledge so gained in a useful form, instantaneously, anywhere in the world. Moreover, it revealed that the United States can act on that information. The combination of new and advanced technologies (C^4ISR, navigation aids, airframes and power plants, stealth, smart weapons, new warheads, decision aids, modeling and simulation, and training systems) provides the basis for seeing and understanding the theater and the battlefield in a new way and acting on that understanding in near real time. As these various technologies improve, air and space power, as its potential was glimpsed in the Gulf, holds out the promise of responsiveness that may be the near equal of continual presence.

This vertical perspective has allowed us to appreciate the potential of the third dimension—the domain of air and space power. This domain is more than a medium through which indirect fires can travel. Properly understood and exploited, it can become a base for operations. Freed from the conceptual limitations of a horizontal perspective, the vertical perspective brings with it the realization that power can now be applied in all three dimensions simultaneously. This potential to effect a transition from the serial or sequential warfare characteristic of the horizontal perspective (even when making use of air and space power) to parallel operations conducted simultaneously and continuously constitutes the essence of the so-called Revolution in Military Affairs.

A New Paradigm for Modern Warfare

This volume does much more than merely affirm the role of modern air and space forces as increasingly essential elements of military power. It asserts that advances in air and space technology offer an opportunity to change our approach to warfare. It argues that emerging capabilities, properly exploited, can free U.S. military strategy and operations from the limits of a horizontal approach to warfare. The classic horizontal approach centers on controlling the intervening terrain between one's own forces and those of the adversary. That approach inevitably reduces the measure of effectiveness of force elements to their contributions to creating favorable conditions at the front. It also ties campaign plans to the requirements to control terrain and to move forces horizontally, over land or seas or both. In this context, it is not surprising that the merits of air and space power are misunderstood and poorly represented in campaign plans, war games, and models.

The implication this study drew from the descriptions of modern air and space forces' core competencies is simply that these forces have overcome the limits of time, space, and mass that dominate horizontal warfare. Central to overcoming those constraints is the ability conferred by space and airborne intelligence, surveillance, and reconnaissance (ISR) capabilities to acquire and operate with a vertical perspective on the theater or the battlefield. This, in turn, not only confers information superiority on the side exploiting air and space power but also allows that side to reconfigure its campaign plan from one based on a horizontal approach to achieving military objectives to an approach that is three-dimensional. The combination of reach, speed, striking power, survivability, and battlespace awareness enables possessors of air and space power to conduct a new type of warfare, unconstrained by the physical and conceptual limitations tied to thinking from a surface combat perspective. In fact, this study argues that land forces and sea forces have increasingly been seeking to exploit the power of air and space to enhance their own styles of warfare.

This study proposes a new paradigm for modern warfare, one centered on domination and exploitation of the vertical domain in order to conduct fully three-dimensional operations. Today, modern air and space power enables the creation of a base for theaterwide operations in the vertical dimension. Unlike bases for ground and naval forces, the bases on which aircraft are deployed and air and space forces operate are not of themselves important to the functions and impacts of air and space power. The value of air and space forces comes from their operations in the vertical dimension, particularly in and over the portions of the battle space in which the enemy operates. In the past, technical limitations meant that air and space forces could not establish continuous control or conduct ongoing operations in the third—the vertical—dimension. Where adversary forces were present, air forces had to fight continually for air superiority to conduct operations. Now air and space forces can attain air and space superiority, even dominance. They can, in effect, exercise air occupation, thereby denying to the adversary any intervening space between the two sides' forces and assets. As a result, air and space forces can establish continuing presence, control, and seamless, around-the-clock operations from the third dimension. This constitutes the attainment of a base for operations.

What is distinct about the idea of a base of operations in the third dimension is that it frees joint forces from the need to engage in a contest to seize or control ground as a means to achieving military and political objectives. Access to the entire theater is available from a position in the third dimension. Coordination of joint-force operations is dominated by the ability to observe and act without the limits and hindrances imposed by geography and adversary forces on forces operating from a horizontal perspective.

Developing a base of operations in the third dimension means concentrating combat power across space and time in a manner designed to pursue one's own strategy while defeating an adversary's efforts to employ its own forces and other assets. The generation of simultaneous operations from the third dimension across the entire theater against all target sets combined with the ability to sustain an extremely rapid tempo of air and space operations can stun, disorganize, and destroy the adversary in as much detail as is necessary to achieve friendly military and political objectives. The creation of a base for operations over the contested theater, involving the seizure of air and space superiority, the establishment of information superiority, the movement of forces into theater under the umbrella of air superiority, and the conduct of large-scale precision strikes will almost without question constitute a de facto defeat of the adversary's strategy. Victory would ensue not simply from the denial of the adversary's objectives but from the successful attainment of one's own campaign objectives and the establishment of a sustainable position of control throughout the battlespace.

From that base, air and space forces—either alone or in conjunction with land and sea forces—can operate in endless combinations anywhere throughout the battlespace in a manner calculated to achieve the objectives of the war. From that base, air forces can exploit the advantages conferred by information superiority, stealth, speed, reach, and precision to strike targets in a flexible manner designed to dominate and, as necessary, defeat the adversary's organization and operations. Continual presence and seamless, around-the-clock operations allow the United States to pursue its objectives regardless of the adversary's behavior; at the extreme, the adversary is rendered irrelevant as a military competitor.

The Future of Air and Space Power

The capabilities of modern air and space forces make the idea of conducting modern military operations without their employment increasingly, even virtually, unthinkable. The kinds of air and space capabilities that are being deployed or that can be created offer the prospect for the conduct of air and space operations in a manner, at a pace, and with an efficiency hitherto unknown. Moreover, these capabilities offer a greater flexibility and freedom of action in the conduct of military operations in the new, and more difficult, strategic environment that is emerging.

The dominant strategic merit of air and space power, as it is or can be employed by all U.S. armed forces, is that it enables a new way of conducting military operations or, as termed by the Strategy Working Group for this study, "a new American way of war." Some of the characteristics of this new way of war are

- the conduct of uninterrupted, seamless operations across the entire theater or battlespace;
- the ability to conduct joint and independent operations in parallel and simultaneously;
- the ability to seize and maintain initiative;
- the ability to dominate the course of hostilities;

- the capacity to deny to an adversary the ability to force an alteration in U.S. strategy and to foreclose its pursuit of strategic alternatives;
- the capacity to defeat adversary forces in the field;
- the provision of "overwatch" or occupation during post-hostilities.

The merits of modern air and space power must be appreciated on three levels. First, there is the level of air and space technology. The increased capability of air and space forces to see, maneuver, and strike targets with great rapidity and efficiency is without question. This technical revolution both enhances the performance of air- and space-oriented capabilities and allows for a significantly enhanced exploitation of the vertical dimension of warfare, thereby moving from a concept of warfare based on the battlefield to one based on battlespace. Second, these new capabilities or performance improvements provide the means to dominate the airspace domain and, from there, increasingly influence the course of events on the Earth's surface. Third, and perhaps most important, the ability to dominate air and space and to operate continuously and fully from that position offers the opportunity to create a new approach to warfare that not only is three-dimensional in character but actually locates the base for operations in the third dimension. From what this study calls the "vertical perspective," the merits of modern air and space power are that they can enable the United States to dominate its adversaries across much of the conflict spectrum. The strategic implication of this new approach is that it permits the conduct of military operations in a manner consistent with U.S. strategy and interests while denying an adversary a viable course of action save cessation of hostilities or defeat.

Concluding Observations

Air and space power can thus serve as the basis for creating a new approach to war. As argued in this volume, air and space power has reached a state of technological and operational maturation that enables it to form the basic building block of a new strategic paradigm. That paradigm would see warfare conducted increasingly from a base for operations located in the third dimension. All forces would use that base for operations to pursue their specific objectives, but not all operations would be located in that third dimension. Three-dimensional warfare presents numerous opportunities to affect an adversary's will and achieve one's own objectives. The overall result would be a national military capability that is comprehensive in character, global in reach, swift in response, and highly effective in its actions.

Notes

1. Stephen A. Cambone, "The RMA and Continental-Size Adversaries: A Case Study" (paper for the Center for Strategic and International Studies, Washington, D.C., January 1995).

2. See, for example, the series of essays in Richard H. Schultze Jr. and Robert L. Pfaltzgraff Jr., eds., *The Future of Air Power in the Aftermath of the Gulf War* (Maxwell AFB: Air University Press, 1992); Richard G. Davis, *Decisive Force: Strategic Bombing in the Gulf War,* Air Force History and Museums Program (Washington, D.C.: GPO, 1996); Colonel David A. Deptula, *Firing For Effect: Change in the Nature of Warfare* (Arlington, Va.: Aerospace Education Foundation, 1995); and Elliot A. Cohen, "The Meaning and the Future of Airpower," *Orbis* (Spring 1995): 189–200.

3. Elliot A. Cohen, *Gulf War Air Power Survey: Summary Report* (Washington, D.C.: GPO, 1993).

Steering Committee

Project Directors:
CSIS
William J. Taylor Jr.
Daniel Gouré

VII Inc.
Charles D. Vollmer
Don W. Rakestraw

Committee Members:

Graham Allison
Harvard University

Stanley Arthur
Lockheed Martin Corporation

Norman R. Augustine
Lockheed Martin Corporation

Jean Betermier
Aerospatiale

Gregory Bradford
Aerospatiale Inc.

Harold Brown
CSIS

Saxby Chambliss
U.S. House of Representatives

Anthony Cordesman
CSIS

Norm Dicks
U.S. House of Representatives

Joseph Draham
EDS

John Egan
Lockheed Martin Corporation

Norman E. Ehlert
USMC (Retired)

Calvin Franklin
Engineering Systems Consultants

Joseph T. Gorman
TRW

Sir Patrick Hine
Air Chief Marshal (Retired), UK

James Hogg
Naval War College

Charles Horner
USAF (Retired)

Fred Iklé
CSIS

Frank Jenkins
SAIC

Phil Merrill
Washingtonian Magazine

Yves Michot
Aerospatiale

Alan R. Mulally
Boeing

Carl E. Mundy
USMC (Retired)

Thomas C. Richards
USAF (Retired)

Brent Scowcroft
Forum for International Policy

George Sibley
EDS

William H. Taft
Fried, Frank, Harris, Shriver & Jacobson

Jerry Tuttle
Mantech International Corporation

James Wade
Defense Group Inc.

Volney Warner
V. F. Warner & Associates

John Weaver
Hughes Aircraft Company

John J. Welch Jr.
CSIS

Larry Welch
Institute for Defense Analysis

David Welp
Texas Instruments

Paul Wolfowitz
SAIS, The Johns Hopkins University

Participation in the Steering Committee of the CSIS project on *Air and Space Power in the New Millennium*, either directly or through a designated representative, should not be construed as a formal endorsement of the project's conclusions and results.

Introduction

Gen. Ronald R. Fogleman, USAF (Ret.)

During this first century of powered flight we have witnessed a transformation of the world as a result of the developments in the capabilities that flow from the medium of air and space. The United States is clearly an aerospace nation. It is almost second nature for us to rely on air and space. Air and space have begun to dominate every aspect of our society, economy, and military affairs.

Despite the preeminence of air power, it is clear that the benefits air power offers the nation are often overlooked. Aerospace products are the number one export of this country. International air travel, spurred primarily by the products coming out of the U.S. aerospace industry, has shrunk our globe. Air power can keep peace-loving people free from war through its deterrent value, its ability to deliver humanitarian aid, and its ability to combine the efforts of allies. It can help shape and drive events around the world.

There are still those who fail to stand back and reflect on the fact that air assets operate in the one medium that surrounds the earth and that touches 100 percent of the earth's population, political capitals, and centers of commerce. Because of the long history of surface warfare and, perhaps, of our very existence on land, air power is not an easy concept to grasp. In 1948, Winston Churchill wrote, "Air power is the most difficult of all forms of military force to measure, or even to express in precise terms." Although there was no TACWAR model back then, Churchill appreciated the difficulty in articulating the value of air power.

This introduction is based on an address by Gen. Ronald R. Fogleman, Chief of Staff, USAF, to the CSIS Final Conference on "The Merits of Air and Space Power in the New Millennium," May 7, 1997, Dirksen Senate Office Building, Washington, D.C.

Today things are changing. Because of congressional caucuses and studies such as this, we are beginning to make a difference. But we must continue to work on the problem. We must understand the way that people look upon air and space, and we must work to shape the ideas that will change that perspective.

Too many people think of air power as simply aircraft engaging in air-to-air combat, or as an extension of artillery for supporting surface forces, or as a protective umbrella to shield surface forces from enemy air power. Clearly, air power plays those roles. But it is far more than that. As our perspective of the world becomes increasingly airborne, through both aircraft and spacecraft, we have in some ways become an aviation-minded nation with a global perspective. In fact, people of all nations are now positioned to appreciate that air power encompasses the ability to totally control and exploit the vertical dimension. Beyond a doubt, this will require a new and robust look, a new assessment, and new understanding of the value of air and space power.

In the face of today's smaller budgets and reduced force structures, it is particularly important to understand the alternatives available when developing strategies about how to maintain a strong, relevant, capable military. Moreover, this process must take into account a world where it is very difficult to predict the future precisely. Those who plan, program, and drive to an absolutely precise future will be precisely wrong. We must, in fact, approach the future through alternatives and understand what we can do to handle them.

Today we talk a lot about revolutions, so much so, in fact, that by now we have become accustomed to being in the middle of them. I am talking not about the revolutions that change regimes and governments but about those that seek to overthrow antiquated beliefs and ideologies.

Faced with the latest revolution—the revolution in information technology—and with the opportunity to see what that means in terms of the future battlefield, or social intercourse, or commerce, the United States Air Force was spurred to an understanding: information capabilities, computing power, and communications links, in combination with air power's traditional characteristics of speed, range, flexibility, and lethality, would lead to a fundamental change in the nature of warfare. That is one revolution that we are dealing with as we try to come to grips with the value of air and space.

The second revolution is in the field of international relations. This revolution was sparked by the collapse of the Berlin Wall, the crumbling of the Warsaw Pact, and the demise of the Soviet Union. We do not need political scientists to tell us that the world is different from what it was a few years ago. The growing number of democratic, free-market societies, combined with the global expansion of information, makes for a world that is more interdependent politically, economically, and militarily.

The third revolution we are dealing with has come to be known as the Revolution in Military Affairs (RMA). The increases in the capabilities of air and space assets as instruments of war have revolutionized our ability to assess and attack adversaries in terms of range, direction, and timing. As an instrument of peace, the RMA has created new expectations for access, influence, presence, and assistance.

The driving force behind this volume is the conviction that air and space power provides our national leaders with *the* best tool for meeting our future security needs. Air and space power possesses attributes of speed, flexibility, range, and precision. Those same attributes make air and space power one of our most adaptable instruments of power in this unstable world—not to the exclusion of our naval, maritime, and land forces, but in conjunction with them.

As air and space power has evolved, however, the difficulties in defining, measuring, and articulating its merits still exist. Recent Department of Defense studies such as the Deep Attack Weapons Mix Study (DAWMS), the Quadrennial Defense Review (QDR), and others tend to emphasize a traditional, linear view of warfare in a two-dimensional world. Although these studies make a sincere attempt to demonstrate the value of air power, they often fail because their analyses concentrate on the tactical utility of air and space power to the exclusion of its strategic and global contribution.

It is easy to quantify the effects of air power at the tactical level; for example, how many trucks and how many tanks are destroyed. These are results we can measure and compare with results from other weapons. It is difficult to show the cause-and-effect of air power when it is used strategically and innovatively.

For one thing, air and space power gives us the ability to look at the battlespace from a new perspective. This comes as a result of advancements in technology that allow us to view top down rather than horizontally. And this top-down approach is the only perspective that is truly theaterwide. This leads to and facilitates the concept of parallel warfare.

I would make the case that Operation Desert Storm was the first conflict in which we were able to accomplish parallel warfare. Air and space power had matured, and the reality of air power finally caught up with the theory we had been developing for so long. Unlike actions during World War II when war was fought sequentially, air power, not the U.S. Air Force but air power as applied by all services, allowed Operation Desert Storm to be fought in parallel.

It is from this perspective that we have attempted to shift toward a new paradigm in warfare. To do otherwise would be a great disservice to our nation. As we recognize that the role and capabilities of air power are changing, we must respond with new ideas and new approaches not just from airmen but from our policymakers, decision-makers, and commanders in chief (CINCs).

It was for that reason that the United States Air Force more than two years ago embarked on a strategic planning effort. The most immediate result of that effort is a document entitled *Global Engagement: A Vision for the 21st century Air Force*. In it we discuss the U.S. Air Force core competencies, which are a means of articulating the contributions by air and space forces in support of our national security strategy. I believe that a coherent vision helps us to focus our strengths and guide us into the future.

Our core competencies—air and space superiority, global attack, rapid global mobility, precision engagement, information superiority, and agile combat support—all offer a theorem for evaluating the merits of air and space power. Further, although these competencies have already been studied in great detail, they are not proprietary to the U.S. Air Force. During our long-range study we looked at the idea that the United States has four air forces, but we quickly came to the conclusion that this is absolutely false.

What we have in the United States is an air force that was established in 1947, formed in the crucible of combat and forged in the fire of the skies over Europe and the Pacific. It was not a benevolent Congress that bestowed independence on the air force in 1947; it was a Congress that had come to realize that the nature of warfare had fundamentally changed as a result of air power's demonstrated capabilities during World War II.

That is why we celebrate 50 years as an air force this year. More important, when the USAF was established as an independent air force, it was given certain chores, certain responsibilities, certain roles, certain missions. The air force is charged with providing capabilities in the air and space arena across the entire spectrum of science and technology, research and development, testing and evaluation, production, fielding, employment, and sustaining of air and space assets.

The other services have air arms—magnificent air arms—but their air arms must fit within their services, each with a fundamentally different focus. So those air arms, when in competition with the primary focus of their services, will often end up on the short end, where the priorities for resources may lead to shortfalls or decisions that are suboptimum. It is therefore important to understand that the core competencies of air and space power are optional for the other services. They can elect to play or not play in that arena. But if the nation is to remain capable and competent in air and space, someone must pay attention across the whole spectrum; that is why there is a U.S. Air Force.

Again, these core competencies are not proprietary; anyone can do any or all. But the U.S. Air Force has no choice; it must be engaged across the spectrum.

That is why, as we worked through the long-range planning effort, we spent much time talking about the idea of air and space superiority and what it means. What we are really talking about is air dominance. Dominance means you own the adversary's airspace. It is not defending your airspace but owning the adversary's airspace—giving him no sanctuary, no freedom of maneuver or of action. Superiority in air and space is consequently at the top of the list of competencies.

We discussed global attack from two dimensions. The first dimension is the ability to reach out from the continental United States while we are mobilizing and moving forces forward. We would not do that in a unilateral fashion; we would do that within some national joint force. Those forces that move forward or strike from the continental United States bases may very well be aided by naval air power coming off a carrier, or by land-based air power, or by any combination.

But we must have the capability to make an impact within the opening hours of a conflict. We have traditionally looked at a paradigm of three phases in a major theater war: the halt phase, the buildup phase, and the counteroffensive. Traditionally, we have seemed to think of our culminating point as being the counteroffensive. But one idea that has come out of the QDR and the new strategy is the idea that a *decisive* halt in the halting phase is critical. This quickly narrows down the adversary's options and expands our options. As options are expanded, the choice can become whether the situation will result in a massive war of attrition or whether the battlefield will be shaped in such a way that our forces engaged on the surface are engaged with immense advantages.

Global attack starts with that long-range capability, but it is complemented by expeditionary fighter forces—the heavy hitters when it comes to generating sorties and combat power. These are the forces that go in and pick up and sustain the air campaign during the later stages of the halt phase and pound and shape the battlefield during the buildup phase.

Global mobility includes sealift as well as airlift. When we talk about rapid global mobility, we acknowledge our shrinking world and those places on the globe where we are called to aid allies, or provide humanitarian assistance, or provide combat forces. We must have air bridges to get there. Therefore we need to recognize and incorporate this capability into the roles that we play. If we are going to be a world power, and if we are going to shape, respond, and prepare, we must have a global mobility capability. It must be a core competence.

When we consider precision engagement as one of our core competencies, we consider far more than the lethal application of air power. Precision engagement can be lethal or nonlethal; it can be kinetic or nonkinetic. It can, for instance, be something akin to a computer network attack on someone's command and control system—the destruction of a critical node. It can be a special operations force that goes in and performs a mission associated with nonproliferation. Or it can be a joint direct attack munition or a laser-guided bomb. It can be any one of these things. But it is going to continue to shape the battlefield. Precision engagement, lethal or nonlethal, kinetic or nonkinetic, is important.

To be successful in the twenty-first century, we are going to have to rely more on information. We must be able to protect ours and exploit the adversary's. Information superiority is not a proprietary competence of the U.S. Air Force, but it is one we must pay attention to.

We have already moved to build impressive offensive and defensive information capabilities, but we must do more. Our responsibilities for disseminating information for battle management and command and control will enable and better integrate our joint forces.

Our responsibility is to provide the joint forces commanders of the twenty-first century with pictures of the entire battlespace—an air picture, a sea picture, and a land picture. This will facilitate the near real-time control and execution of not only air and space missions but also the methods by which we deploy and employ on the battlefield. We have come a tremendous distance in this already, but we have some distance to go.

Air and space power is dependent on, and will remain dependent on, a myriad of support activities. We are an expeditionary force by our very nature. We must be able to pick up, move, go, sit down, and build up forces, and we must do this with forces that are light and agile. Because we must stay focused on this, agile combat support is the sixth of our core competencies.

The U.S. Air Force has recently completed its vision statement, which has been in the field for approximately five months. We have a long-range plan that supports this vision statement, and a doctrine manual that explains how we will fight will also come out to address this statement. This exercise, which began in February 1995, has proved to be extremely valuable for us as a service as we have engaged in the QDR.

We are pleased with the decisions that are emerging from the QDR. We had a vision, we fought hard, and the force restructuring that we are doing has supported where we want to go with our air and space assets and with our people. We feel comfortable. Like everyone else, we would have enjoyed having had more. But more is not in the equation. What we have to do is understand what each piece contributes and then shape it to make those contributions.

This process was clearly assisted by the publication last year of *Joint Vision 2010* by the chairman of the Joint Chiefs of Staff. As a service chief, I have been struggling with the idea that the service chiefs organize, train, and equip forces that we in turn give to CINCs, who employ them in pursuit of our national interests. But no one told us how the CINC would employ forces. Back when the Goldwater-Nichols Act was put into effect, we discovered that the CINCs defined jointness on the basis of the color of the uniform they wore. Just to be safe, they would never do anything unless they had one soldier, one sailor, one airman, one marine; if they had that, they could declare themselves joint.

When Gen. John Shalikashvili, chairman of the Joint Chiefs of Staff, published *Joint Vision 2010*, he gave the whole defense community a construct, four operational concepts—dominant maneuver, precision engagement, full dimensional force protection, and focused logistics. Our exploitation of these concepts along with our information superiority will lead to battlespace dominance. We were able to marry that construct very well with our core competencies and, in the end, validate what we believed the U.S. Air Force would contribute to the national defense equation in the first quarter of the twenty-first century.

We also used another construct, this one from the White House, saying that we have *A National Security Strategy of Engagement and Enlargement*. We have a national military strategy that states simply that we need to prevent, deter, and defeat. From that we have come down to a QDR that states we must be able to shape, respond, and prepare for the future. What we see emerging now is that the Quadrennial Defense Review is not an end in itself but, rather, the beginning of the debate that will shape the future of our military forces.

1

The Coming of Age of Air and Space Power

Daniel Gouré and Stephen A. Cambone

The Cold War is now almost a decade behind us. A new millennium is immediately before us. The United States is faced with the necessity to assess anew its national security strategy and, in that context, to decide its requirements for military power. This will not be an easy task. There is no great threat to the nation's survival or well-being. Having but recently emerged from a 40-year global struggle to defend the community of democratic and free market nations, the American people are hungry for peace and for what some have called a peace dividend. Numerous challenges at home cry out for attention and resources. As a result, there are demands for change in the nation's security posture and the structure of its armed forces. There is no consensus, however, about the objectives of that security strategy nor about the character and purposes of American military power.

We are just now beginning what is likely to be an extended period of debate on these issues. This debate should be informed about the changing nature of military power in the twenty-first century brought on by the evolution of technology and the new ways that military power can be applied in peace, crisis and war. The Department of Defense (DOD) has just completed the congressionally mandated Quadrennial Defense Review (QDR) intended to assess the adequacy of U.S. national security strategy and capabilities to meet the threats of the next decades. The QDR will be reviewed and critiqued by the National Defense Panel (NDP). It is doubtful that either the QDR or its companion effort, the NDP, will answer fully the concerns that have prompted this debate.

This project of the Center for Strategic and International Studies (CSIS), *Air and Space Power in the New Millennium*, is intended to assist the efforts of the administration and Congress to answer the pressing questions before them. But why conduct a study on air and space power? Air and space power constitutes one of the strongest elements of U.S. military power: it is a demonstrably unique national competence that provides an asymmetric advantage over potential adversaries. The role of air and space power in diplomacy, crisis, and war has changed and expanded dramatically over the past century, as air and space forces have developed the capability to perform more missions, reach out farther, and strike with more force. How, then, in an era of changing national security concerns when decreasing resources are devoted to defense, should the nation understand the contributions of air and space power within the restructured armed forces, not only in meeting current security requirements, but also in shaping a U.S. interests-based security environment for the decades ahead?

The QDR and NDP are taking place amid a reassessment of the maturity of air power. There is no question that air and space power comprises an important element of modern military strategy and capabilities. It is the character of that power and the extent to which it can be made central to strategic and operational planning that separate proponents and skeptics. Does air power have an independent role? Is it now time for the services, including the United States Air Force (USAF), to transform their doctrine and force structure into a new concept of warfare? Can we transform warfare? Lessen the bloodiness of "direct contact"? Enhance our capacity to prevent or deter hostile actions through the exploitation of air and space forces? We have been debating these questions for nearly 50 years. Until now the answer has been no.

This study was initiated to assess whether that answer should be reconsidered and whether the prior claims for air power now rest on a sound footing. If they do, air and space power, given its potential, ought to be treated not as another element of the restructured force, but as its fundamental building block. That the historical answer may need to be revised is reflected in the study's working definition of modern air and space power: air and space power is the ability to conduct military operations in three dimensions, simultaneously and globally, from a base of operations in the third dimension—air and space. In strategic terms, it requires the ability to control and exploit the air and space medium. This, in turn, requires the ability to deploy, command, control, and operate effective offensive, defensive, and movement capabilities in and through the air and space mediums while denying, as necessary, those capabilities to an adversary. Modern air and space power consists of the following:

- space systems, including the means for placing objects in space;
- air systems, both manned and unmanned fixed-wing and rotary platforms;
- missile systems operating from above, on, or—in the case of underwater platforms—below the surface of the earth against targets in air, in space, or on the surface; and
- the command, control, communications, computers, intelligence, surveillance, and reconnaissance (C^4ISR) systems that enable linkage to all of the above systems.

To address the merits of modern air and space power, CSIS brought together more than 150 experts from the military, industrial and academic communities. A steering committee was created to guide and advise the study, consisting of retired senior flag-rank officers from each of the armed services, leaders in the defense industry, and senior experts from academic and analytic institutions. Working groups were formed to address a strategical framework for modern air and space power:

- strategy
- air and space superiority
- global attack and precision strike
- information superiority
- technology and the industrial base
- rapid global mobility/agile combat support

The members of the working groups included individuals who had served in each of the armed services, the executive office, the Departments of Defense and State, and other government agencies involved in national security affairs. Each of the armed services offered valuable assistance through high-level briefings on their programs relating to air and space power.

The Emerging Security Environment

The potential that modern air and space power offers the nation is particularly significant in light of the changing strategic environment and its demands on the character and employment of U.S. military power. The United States no longer faces a direct threat to its own survival. It is confronted, however, with a disorderly, some would say unstable, world and a set of complex, evolving geostrategic relationships that do not provide the impetus for a rapid focusing of attention and swift responses.

The nation is currently the beneficiary of its Cold War efforts with respect to military doctrine and strategy, forces and technology, and alliances and forward deployments. This legacy is a dwindling asset, subject to the effects of declining defense budgets, technological obsolescence, force reduction, and overseas base closures. In addition, the strategic, political, and operational circumstances in which we can expect to employ our military forces are changing in the aftermath of the Cold War and are likely to continue to evolve in the foreseeable future. As a result, the United States must restructure its military strategy and forces in line with clear budgetary trends, the evolving nature of the international environment, and the inevitable advance of technology.

In this context, it is a mistake to focus U.S. military strategy on two "canonical" regional contingencies that have served as the basis for force planning over the past four years. A changing strategic and political environment makes it difficult to know where or when the next regional crisis will arise. The two theaters against which the United States has focused its planning, Northeast Asia and Southwest Asia, are unusually favorable; both are mature theaters that offer extensive infrastructure, the presence of U.S. forces on the ground, relatively lenient rules of engagement, and well-rehearsed plans for mobilization and response in the event of war. Moreover, neither of these scenarios reflects the potential robustness of the adversaries who might confront us in the next century.

The adequacy of U.S. forces to support national security and military strategies must be evaluated relative to a range of challenges to U.S. security and global interests that probably will be more difficult to meet than those we have seen since the Cold War ended. In addition, the context of U.S. involvement is likely to be different from that of the Cold War, when U.S. forces were forward deployed, defending well-established borders. Finally, potential U.S. adversaries are clearly not standing still. They are taking advantage of opportunities presented in the international arms market.

It seems increasingly probable that the future international security environment will be more complex and stressing than the one we confront today. Whereas current scenarios envision opponents to be rather limited in size, capabilities, and reach, this may not be the case in the future. The QDR's vision of the future security environment suggests the need to prepare to meet the threat of robust regional adversaries early in the next century and the prospect of heavily armed, theater-level "peer" competitors or major power by approximately the year 2014. Although these major powers are unlikely to have the means to pose a global challenge to the United States, they will have sufficient capability to compete with us and our allies within their own theaters of influence. In that context, they will constitute theater-level peers that potentially approach a "continental-size adversary."[1] In World War II, neither Germany nor Japan was a global power. In the end, neither had sufficient strength to win against the weight of the combined forces of the Allies. Yet both conducted continental-size wars for a period of years. Theater-level peers will have sufficient resources and occupy enough territory to become resilient adversaries. In addition, they are likely to possess weapons of mass destruction (WMD) and the long-range delivery systems needed to threaten distant U.S. interests and even the U.S. homeland. They may be members of an alliance structure creating the possibility that they too can mobilize coalitions to oppose a U.S.-led coalition.

Future adversaries, whatever their "size," will no doubt be tougher politically, operationally, and technologically than they have been in the past. They probably will possess WMD and delivery systems in militarily significant numbers and configurations. They will employ their forces jointly and have access to C^{4}ISR capabilities better than much of what the U.S. deployed only a few years ago in the Gulf War. Presently, potential opponents are attempting, some more successfully than others, to acquire advanced military capabilities that could pose a serious challenge to the United States. Among the advanced capabilities that will be of greatest concern for U.S. in-theater forces are airborne and space-based surveillance, communications, and navigation and targeting systems coupled to long-range delivery systems (such as tactical ballistic missiles and cruise missiles). Many of these capabilities are or soon will be available in the commercial or military markets.

Regional or theater-peer competitors need not build military forces symmetrical to those of the United States to mount a significant challenge. In many cases, they need only to focus on denying or minimizing the U.S. forward presence and the ability of the United States to intervene in their region. Many potential regional and theater-peer competitors are acquiring the capabilities necessary to complicate any U.S. intervention. Among the capabilities actively being procured, in addition to remote sensing and long-range surface-to-surface strike assets, are the following platforms: quiet diesel submarines, advanced sea mines, anti-ship cruise missiles, modern combat aircraft of more advanced designs armed with high-performance air-to-air weapons, integrated ground-based air defenses, and long-range artillery. Adversaries can also exploit the regional environment to create opportunities for unconventional attacks on U.S. forces or critical targets. In those scenarios where the United States is not already forward-deployed, anti-access strategies based on these capabilities could severely degrade U.S. power projection capabilities. In addition, potential adversaries can take other defensive measures to limit the effectiveness of U.S. power projection. Camouflage, dispersal, and, more important, burying and hardening of targets are options that have been employed by U.S. regional adversaries in the past.

In addition to the prospects for theater-peer competitors and robust regional adversaries, the United States will encounter challenges in lesser contingencies. These contingencies will impose a need for advanced capabilities to meet challenges that are stressing to commanders and political leaders seeking to minimize both committed numbers of forces and casualties. Against this desire to deploy limited forces for limited aims, adversaries could prove to be less constrained either in numbers or in the range of threats they present. Adversaries can oppose the U.S. force directly or interpose obstacles (real or virtual) between the force and its objectives.

Without reliance on other force elements, modern air and space power cannot be expected to meet all potential threats or challenges. In many lesser contingencies, primary combat power will come from forces on the ground. Yet there is almost always some element of air and space power present in support of modern military forces wherever they are deployed. In many cases, problems that seem immune to an air power solution—drug smuggling, human rights violations, environmental degradation, even terrorism—are relatively immune to any military solution.

It is wrong to make the test of the relevance of modern air and space power to national security its relevance across the entire conflict spectrum. What is important is that it can address the broadest portion of that spectrum and do so in a manner that can increase expectations that the United States will be able to achieve its security and defense objectives with the least risk. Air and space power appears to offer increased response options to decision-makers and military leaders confronted by an uncertain threat, changing political and military conditions in zones of conflict, the need to wait until the adversary makes the first move, and the requirement to project power rapidly over long distances into theaters of widely varying character. It should be noted, however, that the ability of modern air and space systems to meet the challenges posed by military threats of widely differing character and intensity suggests that some of those same capabilities—or new systems derived from them—should be capable of addressing requirements posed by many of the so-called new threats.

The Merits of Air and Space Power

Air power has been a factor in warfare since World War I. Its potential was studied by all the major powers in the interwar period. During that time, much of the doctrinal thought devoted to air power sought ways of breaking the tactical deadlock experienced during World War I.[2] Some nations, such as Great Britain, saw enough merit in air power to create an independent air arm and to pursue military strategies based on the idea that air power could constitute the principal element in future conflicts.[3] Air power advocates believed that the ability to overfly opposing land armies and strike an adversary's vulnerable strategic rear, if done with sufficient power, would offer a new means for imposing one nation's will on another.[4]

In World War II, air power became more than an adjunct to the signal corps and a home to modern knights. It proved to be an essential feature of modern warfare. As perhaps the most authoritative source on the role of air power in this period noted, "in this war, air power may be said to have reached a stage of full adolescence."[5] The sheer magnitude of air operations in World War II was evidence of this. In the European theater alone, the Allies dropped over 2.7 million tons of bombs, conducted over 4 million sorties, and lost more than 40,000 planes and nearly 160,000 men.[6] World War II also clearly demonstrated, however, air power's continued lack of maturity. Although it was essential to the war initially, and at times decisive at the tactical and operational level (for example, in supporting the Normandy invasion and the subsequent breakout), the ability of air power to achieve decisive strategic results independent of land and naval forces was ultimately disappointing or questionable at best.[7]

In the ensuing decades, with the advent of nuclear weapons, ballistic missiles, and security strategies based on the principles of deterrence, air power continued to mature and the idea of an independent air operation came to have a prominent place in thinking about the role of air power, and, soon after, space power. Nevertheless, it was at the tactical level that air power made its most visible contributions. Improvements in aircraft technology increased the role of air power in the support of ground forces. The fielding of tactical missiles and rotary wing systems further amplified the already considerable role of air power in the provision of remote fires on the tactical battlefield. U.S. air power proved to have a flexibility, responsiveness, and effectiveness that were critical to tactical operations from Korea to Vietnam.[8]

The 1991 Gulf War, which demonstrated the power of new technologies (stealth, precision guidance, airborne surveillance, space-based sensors and communications) and the capacity available in modern battle management and command, control, and communication (C^3) systems to control and direct them, sparked a debate as to whether air power had reached maturity and whether the "coming of age of air and space power" meant that the use of independent air operations to achieve strategic objectives was a realistic prospect.[9] The ability to control the skies and to deliver precision ordnance with impunity virtually anywhere in the theater marked a qualitative leap in the capability of modern air power. Although precision weapons constituted only 9 percent of the total tonnage dropped during the Gulf War, they accounted for more than 42 percent of the damage to strategic targets.[10] A small number of precision-bombing-capable aircraft also inflicted a disproportionately great number of casualties on Iraqi field forces.

The complete realization of air and space power's strategic potential is possible today because, for the first time, air and space power can deliver almost fully on the vision put forward by the early air power advocates. Historically, technical limitations in target identification, navigation, and weapons delivery often meant that even massive air operations produced few tangible results. These limitations often resulted in the need to mass platforms over the target. Given the relative strength of anti-air defenses, the inherent vulnerability of air platforms, and the requirement to fight to and from the target when air superiority was not secured, the utilization of large numbers of aircraft often created extremely high losses in strategic bombing missions. The synergy of improved intelligence and surveillance, accurate navigation, precision targeting and weapons delivery, and reliable, real-time communications greatly improved survivability, and the availability of a variety of delivery means has fundamentally changed the character of warfare in and from air and space.

As a result of improvements in technology, not only can air and space forces perform their missions under ideal conditions, they also can increasingly operate in adverse circumstances. Many of the historical limits on military operations, particularly those that involve air forces, are being overcome. No longer does nighttime provide a cover for adversary movements. Moreover, ongoing technology development efforts may significantly reduce the protection from detection and attack afforded by foliage and camouflage. Weather, possibly the major impediment to air operations during the Gulf War, may be amenable to a variety of technology solutions. The provision for global positioning system (GPS) guidance for aircraft and munitions is easing considerably the problem historically posed by weather to effective air strikes.

Modern air and space power consists of a number of critical attributes, what the U.S. Air Force has termed "core competencies." These attributes are core competencies both because they are essential to virtually any air operation, as well as to modern joint operations, *and* because embodied within them are system capabilities that are such an improvement over preceding generations that they are the basis for a quantum leap in the overall effectiveness of air and space power and, thus, of U.S. military forces. The newly acquired capabilities of air and space systems to perform substantially better than prior generations and to leverage the capabilities and advantages of other systems and forces constitute a set of merits for air and space power. In different ways and to somewhat different degrees, each of these is also an important, even central, feature of the future war-fighting capabilities of all U.S. military forces. These core competencies or critical attributes of modern air and space power are

- information superiority,
- global attack,
- precision engagement,

- air and space superiority,
- rapid global mobility,
- and agile combat support.

Information Superiority

The core or backbone of modern warfare is information.[11] As pointed out by the Joint Chiefs of Staff in *Joint Vision 2010*, information superiority (IS) is the sine qua non of desired U.S. military capabilities. Common battlespace awareness, location certainty for friendly and adversary forces, and real-time dissemination of information create the basis for seamless, joint operations. The instantaneous and essentially cost-free collection, processing, and distribution of information can also change the tempo of operations, providing for a higher degree of flexibility and responsiveness than heretofore existed. Information superiority is vital to the ability to employ precision strike assets in a manner most likely to create the desired results.

Improvements in navigation and position location constitute an important change in air and space power. For air forces, finding the target was often a difficult task. For ground forces, knowing the position of one's own units was often more important than being able to locate opposing forces. With modern navigation aids, particularly GPS and similar systems, navigation is no longer the problem it once was for any force. As a result, there is now an ability to know both the positions of one's own forces and those of the adversary. The application of precision location technology to weapons delivery has further raised the probability that a weapon can be placed on a target with confidence.

Additionally, communications have become more important as information collection has approached real time and as weapon systems have acquired the ability to respond to such information. Real-time communications can tie global capabilities simultaneously to multiple sets of specific forces, performing particular missions, at a local place and time. They also serve to lessen the importance of time and distance as factors in modern combat.

Air and space power provides the ability to see the entire theater/battlespace in three dimensions and to use the information gained to develop an appreciation for how an adversary performs as a complex system—in other words, study his ability to react to stimuli and to learn. With this information, air and space power allows force—stimuli—to be exerted at critical points within the adversary's system *very nearly as they appear*. The object of military operations shifts from "crossing the deadly ground" to destroying those parts of the adversary's system on which his resistance depends at the moment and in the place best calculated to bring about his defeat. In this way, air and space power is applied as an independent force to achieve strategic results.

Achieving information superiority will have wide-ranging effects on the way that American forces plan and conduct military operations. Common battlespace awareness and assured real-time information will allow for a greatly accelerated tempo of combat as well as highly accurate identification of friendly and hostile forces on the battlefield. A streamlined, highly networked information dissemination and retrieval system will allow for greater latitude in tactical combat because tactical units will have timely, relevant information on the battlefield. Real-time sensor to shooter coupling, enabled by wide band links and intelligent data bases, will reduce the number of critical nodes manned by humans. The concentration of fires enabled by information-enriched weapons will make the battlespace more lethal and, necessarily, more dispersed. As a result, military forces and their support structures (including command and logistics) must be far less concentrated than ever before if they are to survive and operate. This implies the potential for a reduction in numbers of platforms (many of which will be unmanned) and a concomitant increase in the information content of remaining platforms. Precise long-range fires can be delivered from remote air- , land- , or sea-based platforms. Finally, because of these qualitative changes, the eternal balance between centralized command and decentralized execution will be more important—and even more difficult to achieve.

Information warfare (IW) will be the means whereby information superiority is achieved. IW has both offensive and defensive applications and can be applied at the tactical, operational, and strategic levels of conflict. Strategic IW includes the targeting (offensive) or protection (defensive) of homeland assets such as information and inter-netted critical national infrastructures. Operational IW includes the ability to target or protect information assets at the level of campaigns. At the tactical level, IW entails the ability to disrupt, disable, deny, or exploit enemy battlespace information capabilities and protect one's own.

Protecting our own critical systems is essential to maintaining information superiority. Protecting systems and information requires strong measures for defensive IW: strong encryption, intelligent firewalls, and other countermeasures, integrated with such traditional skills as personnel and physical security. Offensive IW must include kinetic attack capabilities directed at critical infrastructure nodes (as shown during Operation Desert Storm), as well as digital attack capabilities to enhance the traditional electronic warfare mission (jamming and exploiting adversary signals).

Finally, information superiority, promises to enhance the effectiveness of the joint force. Seamless interoperability of heterogeneous systems across and within services is a precondition to the achievement of IS. A common, consistent awareness of the battlespace would promote the synchronization of forces and maximize the use of military assets across services to accomplish goals and exercise a decisive advantage. The overriding goal here is to end the spatial boundaries and "turf" that formerly marked off the battlespace among the services.

Although a wide variety of systems contribute to the attainment of information superiority, airborne and space-based systems are at the core of the emerging U.S. capability to dominate the battle for information in future conflicts and crises. With new sensor, fusion, and dissemination technologies, and the advent of airborne command and control, aerospace forces have expanded their competencies to cover the entire C^4ISR mission. The U.S. Air Force's Rivet Joint electronic signals monitoring, airborne warning and control systems (AWACS), and Joint Surveillance Target Attack Radar System (JSTARS) have proven themselves to be essential platforms in conflict areas. Information gathering and disseminating unmanned aerial vehicles (UAVs) such as Predator, Global Hawk, and Dark Star are being developed. The use of tactical UAVs is already occurring in the United States Navy and is under consideration by the United States Army and United States Marine Corps. Spaceborne defense reconnaissance, communications, and geolocation capabilities are vital to U.S. military operations at all levels. These capabilities will only improve as new adaptive optics satellite technologies with ever-improving resolution are deployed. All of these systems will be supported by highly fused knowledge-bases and worldwide data dissemination systems. Each of these air- and space-based assets is a fundamental hardware component of—and provides content products for—an emerging "system-of-systems" that will facilitate twenty-first-century information superiority: the global command and control system (GCCS) coupled with the global broadcast system (GBS). The ability of other services to do long-range remote fires rests on the availability of USAF information-gathering and dissemination capabilities.

Global Attack and Precision Strike

The United States, alone among the nations of the world in its employment of air and space power, has the capacity to act globally—meaning anywhere. Information superiority increasingly provides the ability not only to see the world anew, but also to view it globally—meaning comprehensively. The third element in the ability to act globally—meaning over a long range—rests on long-range delivery systems that can reach virtually anywhere. U.S. air- and sea-based global and precision strike capabilities provide the means for timely response to aggression anywhere on the face of the earth. Such capabilities can be critical to seizing the initiative in conflict by taking early action against adversary forces in order to halt an offensive, neutralize long-range delivery systems and WMD, dismantle air and coastal defenses, disrupt command and control, and interfere with other military activities. Additionally, such capabilities offer the opportunity for use at less-than-maximum range. When deployed into the theater but operated from just beyond the reach of any adversary, global strike capabilities can achieve at a relatively high sortie generation rate, thereby providing large amounts of firepower on a timely basis.

Only air and space forces have the range, speed, deployability, and flexibility to engage rapidly, survivably, and accurately in any region. The unique contributions of air and space forces to global attack and precision strike are their capabilities to provide a quick and tailored response to areas essentially unreachable by surface forces. For example, when used in response to a cross-border invasion in a short-warning scenario, long-range air power equipped with smart munitions has the capability to halt enemy movement before critical territory is lost.

Perhaps more important than the timeliness provided by globally available attack capabilities is their certainty of response. The ability to see the battlespace and to have knowledge of what is occurring provides only part of the answer. The other part is the ability to strike anywhere, regardless of the adversary's efforts to deny that ability. Long-range bombers—B-2s armed with precision-guided munitions (PGMs) and B-1s and B-52s with standoff weapons—and cruise missiles offer the prospect of certain action. In addition, U.S. Navy ships and aircraft carriers provide a companion, equally important capability for global attack. Carrier-based aircraft operating in littoral waters, in effect, can project power halfway around the world and hundreds of miles inland. Sea-launched cruise missiles can reach even deeper, up to 900 miles from the launching vessel.

The United States also maintains, and will unquestionably do so for the foreseeable future, a global attack capability centered on strategic nuclear forces. Although the number of strategic systems has declined, under START II limits the United States will maintain a triad of some 3,500 warheads. Under the proposed principles for START III, we will reduce that number to 2,000–2,500 warheads. This capability is likely to remain central to U.S. national security strategy for the foreseeable future not only to deter threats to the U.S. homeland but also, if necessary, to counter regional WMD threats that cannot be addressed by conventional offensive and defensive means.

Precision strike is much more than a relatively high probability of hitting targets. Precision strike as a force capability is closely linked to the attainment of air, space, and information superiority. To be effective, precision strike operations rely on a high expectation of platform survivability. In addition, they are a necessary complement to defensive actions against an adversary's long-range threat (e.g., missiles and aircraft) and are a vital part of a suppression of enemy air defense (SEAD) campaign to achieve air superiority.

The impact of the ability to conduct precision strikes is most significant at the operational or campaign level. Properly supported, precision strike provides massed effects without the need to mass forces. The capability for one pass or mission to achieve one or many kills not only provides efficiency, but also frees up assets (platforms) to cover more targets, thereby permitting either breadth of coverage or intensity of firepower as needed. This creates the opportunity to conduct a precision strike operation that takes advantage of the ability to see and understand the battlespace. It can focus on simultaneously disabling and dismembering an adversary through focused attacks.

At the operational level, the unique properties of global attack and precision strike allow U.S. air and space forces to shape the battlespace from a distance—to significantly influence the adversary's area of operation from outside his area and beyond his reach—thereby minimizing the placement of friendly forces in harm's way. The fewer U.S. forces placed at risk, the more credible U.S. commitments and the more effective operational employment. Our freedom of action creates operational opportunities while eliminating the adversary's options.

Overall, the power of a precision strike campaign is its ability to more efficiently exploit air and space power in the pursuit of operational and campaign objectives. This ability supports the campaign objectives of seizing control from the adversary and shaping his behavior. The ability to gain and maintain control can, in turn, reduce the duration of a conflict. The efficient application of air and space power is vital in an era of relative asset scarcity, with the possibility of near-simultaneous conflicts in widely separated regions and the threat of weapons of mass destruction in any theater of war.

Precision strikes also present opportunities to counter a responsive threat—one that would meet advancing air defense systems or otherwise protect targets from attack—or to address the problem of striking targets where there is otherwise a significant risk of collateral damage, such as in an urban environment. Defensive responses to increasing U.S. capabilities to control the air and to conduct remote fires include hardening or burying targets, deploying mobile targets near or in high-value or sensitive locations such as hospitals and schools, and increasing the mobility of targets. Precision strike capabilities increase the likelihood that targets can be struck as required with adequate probability of destruction and a minimum risk of collateral damage.

With the advent of stealth, precision navigation, standoff delivery, electronic countermeasures (ECMs), and computer-based mission planning using digital information, the dilemma of balancing effectiveness against survivability has been solved. Historically, effectiveness required that platforms go "lower and slower," and survivability urged "higher and faster." Today, concerns for survivability need not be satisfied at the expense of effectiveness. The operations of F-117s and cruise missiles in the Persian Gulf demonstrate very clearly that modern air power allows effectiveness to be coupled with survivability.

The combination of these long-sought-after but only recently attained capabilities translates into greater reach, agility, control, and targeting potential for forces exploiting modern air and space means. Such modern air and space systems, supported with a combination of speed, stealth, and new weapons, constitute a much deadlier and more effective force than has been available in the past, not only to air forces, but to land and naval forces as well.

This qualitatively new level of functionality allows air forces to achieve the effects of mass without the need to mass forces on the scale required for wars fought earlier in the century. The statistical basis for this claim is well known and demonstrated. Whereas in World War II it took 3,000 aircraft and 9,000 bombs to provide 90 percent assurance of destroying a target, it now takes as few as two aircraft and four bombs.[12] Beyond the obvious increase in efficiency in the delivery of ordnance per platform made possible by the precision delivery of weapons, the attacker is now able to employ a smaller fraction of his available power to achieve disproportionate results or to spread these results over a far wider area. In a recent series of tests, the U.S. Air Force demonstrated the B-2's ability to deliver up to 16 independently targetable weapons in a single pass or mission. As a result, planners can now distribute available power far more effectively than when they had to depend on mass to achieve the desired effects.

The ability to deploy one's own forces more effectively and across a wider target set creates the opportunity for precision strike campaigns. The breakthrough in precision targeting is less "one pass, one kill" than the potential to conduct many discrete passes with discrete kills simultaneously across the entire theater battlespace. Consequently, rather than having to concentrate his forces against sets of high-priority targets in series, U.S. air and space forces can use smaller numbers of systems to cover greater numbers of targets simultaneously. These factors, coupled with the inherent reach of modern air and space platforms (in the case of aircraft with refueling), mean that, at the limit, air and space power, from the attacker's perspective, can *be* anywhere, and from the adversary's perspective, *is* everywhere.

In addition, the ability to conduct precision strikes offers prospects for changing the character of air operations even in limited conflicts. U.S. forces are confronted with the need to carefully tailor their use of force in so-called operations-other-than war. In the past, constraints on the use of force existed because of uncertainties about operational success and resulting effects. Precision strike capabilities increase the probability and predictability of success and decrease the prospects of casualties and collateral damage. Modern precision strike assets provide the means to employ air power in support of other forces under rules of engagement that impose constraints where the prospects of collateral damage exist. During operations against Serbian forces in Bosnia, the United States engaged in its first precision strike operation. Approximately 70 percent of the munitions used by NATO forces were precision weapons. This ability to successfully use the overwhelming advantage that NATO forces had in air power may be instructive regarding the utility and usability of air power in future operations-other-than-war (OOTW).

Air and Space Superiority

Air and space superiority is an essential component of modern warfare. U.S. military forces and operations are increasingly oriented around the attainment and/or exploitation of information superiority, mobility, global reach, and precision strike. Each of these capabilities requires that the United States have the necessary degree of control over the air and space domains, and that U.S. forces, at a minimum, can deploy and function without significant interference. At the same time, the United States needs to deny that freedom of action to any adversary. The strategic impact of air and space superiority is realized when the degree of superiority allows us to pursue our preferred strategy while simultaneously denying that option to the adversary.

Throughout history, the attainment of air superiority has been critical to strategic and operational mobility by all forces. Germany's inability to control the skies over northwestern Europe inhibited the Germans from attacking England in 1940 and made possible a cross-channel attack by the Allies some four years later. The absence of air superiority has been keenly felt by twentieth-century armies, reliant as they are on maneuver and vulnerable as they can be to interruptions of supply. The advantage afforded to a military force that is both able to maneuver and has control over the air and space is almost incalculable. The effects of the loss of control over those dominions, however, has been demonstrated repeatedly in warfare from the Falaise pocket in 1944 to the Mitla Pass in 1967 and northern Kuwait in 1991.

Air and space superiority offers the national political leadership the freedom to engage globally at any time and in any place—in sum, the freedom to exercise national prerogatives. It is the critical, synergistic enabler for all forms of military power, ranging from other air power projection forces to ground and naval forces. With air and space superiority, the United States can use the full range of its air power forces to maximum effect, including ones that might otherwise be vulnerable to an enemy contesting control of the skies. Air and space superiority reduces losses and markedly increases the effectiveness of friendly military operations. Overall, control of air and space provides both the freedom to attack and—in many ways more important—freedom from effective counterattack.

The task of achieving air and space superiority has become more complex over time. Historically, it was principally a problem of air forces versus air forces. Over the past several decades, the attainment of air and space superiority by modern air forces has involved a battle between air forces and ground-based air defenses. This was first and most graphically demonstrated during the 1973 Middle East War. It was reconfirmed during the 1991 Gulf War in which less than 1 percent of the coalition aircraft losses could be attributed to hostile air activity, compared to nearly 90 percent from ground-based missiles and antiaircraft artillery.[13] It must be expected that the battle for air superiority will increasingly involve early and, probably, massive strikes against a range of ground targets (e.g., air defense command centers, communications, radars, surface-to-air missile [SAM] batteries, airfields), further blurring the distinction between the air campaign and the overall theater campaign.

Space superiority is most often associated with the ability to reliably launch platforms into space and, once they are there, to operate them effectively. The maintenance of reliable access to space and the exploitation of the space domain to ensure information superiority and robust communications are unquestionably major parts of ensuring space superiority. Another is the protection of space-based assets and links from interference or attack. As more nations acquire advanced electronic warfare and even space-launch capabilities, the vulnerability of key assets and capabilities needs to be addressed. Finally, the ability to gain, maintain, and exploit space superiority must also include the capability to operate effectively against hostile assets in space or elsewhere in the three-dimensional battlespace.

Whereas achieving air superiority is increasingly well understood, space superiority may take on a slightly different definition. Space superiority provides the ability to exploit space as a critical integrator and enables a range of operations. Unlike air superiority, however, space superiority can be achieved and maintained in peacetime. The critical issue is, can space superiority be retained in time of conflict? Electronic countermeasures and increasingly sophisticated information operations, using technology that proliferates through such systems as the Internet, can potentially have an impact on the United States's ability to retain space superiority in war. Commercial satellites and the widespread development of communications, navigation, and weather satellite systems by many nations ensure that the information and support generated by space systems will be available to even the poorest nations or threat groups. Destroying these systems to deny their availability to the enemy would have the effect of alienating nations or international commercial or treaty concerns that might not otherwise be involved in the conflict. Thus, assuring space superiority in battles of the future will require whole new technologies and concepts and a diplomatic framework that will protect U.S. and allied systems while selectively denying the capabilities of commercial and third-nation systems to the enemy.

Mobility and Logistics

Mobility and logistics are often overlooked as core competencies and key factors in the operation of all U.S. forces in peace or war. Mobility capabilities impart energy to the system of worldwide U.S. military deployments and activities; they support the entire array of military engagements from regional conflicts to peacekeeping, humanitarian assistance, and conflict prevention. It is obvious that the United States could not conduct itself as a world leader and could not project power into regions of interest without large, robust, and agile mobility and logistics capabilities.

Even when the value of mobility and logistics is recognized, too often they are viewed as relevant only in the space between points of embarkation and debarkation. Modern, three-dimensional warfare will require that mobility and logistics be integrated into the heart of combat and maneuver operations. Exploitation of the opportunities provided by information superiority can allow mobility forces to modify their operations, thereby changing their signatures and footprints to reduce visibility and vulnerability to an enemy. Information superiority will also change the nature of the logistics pipeline from the current bring-everything-and-sort-it-out-in-theater approach to one based on lean logistics and just-in-time logistics.

Mobility capabilities include the tanker support fleet that frees modern aircraft reliance on fixed airfields or aircraft carriers. As a result, the issue of distance has fast been eliminated as a factor in the application of modern air and space power. Air refueling can take U.S. aircraft halfway around the world and back. It allows the deployment and even the dispersal of air assets beyond the range of the adversary's long-range strike systems while permitting their reconcentration as needed to perform a mission. Air refueling is essential to providing near-continuous presence as well as global reach.

The core competencies of air and space power are in themselves merits insofar as they enable air and space forces to operate more effectively and efficiently than before, to leverage and support other forces, and to provide expanded or different military options to decision-makers. It would be a mistake to limit our appreciation of air and space power to the increased capabilities represented by these core competencies individually. Taken as a whole, however, these core competencies offer the United States potentially new and decisive strategic advantages.

Considering the Merits of Air and Space Power

The dominant merit of air and space power, as it is or can be employed by all of the armed forces, is its strategic significance—it enables a new way of conducting military operations. This study's Working Group on Strategy termed this "a new American way of war," characterized by the following:

- the uninterrupted, seamless conduct of operations across the entire theater/battlespace once engagement begins;
- the ability to conduct joint and independent operations in parallel and simultaneously;
- the ability to seize/maintain initiative in a manner that undermines an adversary's strategy, thereby enabling the United States to pursue a strategy of choice;
- the ability to dominate an adversary so as to deny him the ability to force an alteration in U.S. strategy and to foreclose his pursuit of strategic alternatives; and
- the capacity to shape and control a theater of interest through the provision of an "overwatch" or air occupation capability post-hostilities or in lieu of hostile action.

Why pursue a "new American way of war?" There are several reasons. First, such a capability is in keeping with the U.S.

strategy of being actively engaged in the world and committed to the defense of friends and allies around the world. In that context it is desirable to create the kind of military power that can dissuade most potential adversaries from contemplating aggression. Second, the ability to seize the initiative recognizes the likelihood that our adversaries will initiate hostilities. Third, this new way of war offers the prospect for controlling the course of a conflict and shaping the adversary's behavior, eventually rendering him inert and helpless. Recognizing the inevitability that advanced military capabilities, including weapons of mass destruction (WMD), will proliferate into the hands of potential adversaries, it is important that the United States have the capability to control the course of hostilities and shape an adversary's behavior, including the ability to swiftly and decisively negate an adversary's use of advanced conventional and WMD systems that pose a threat to American or friendly forces. This style of warfare offers the greatest opportunities for a rapid and successful termination of hostilities in a manner supportive of U.S. objectives. As a result, the prospect is greater that there will be fewer casualties on both sides and less destruction overall.

The merits of air and space power also should be judged in terms of the utility it affords to and derives from the basic operating "style" of other forces. In particular, the combination of air and land power is what will enable the United States to win wars. Land power poses to adversaries the threat of one form of maneuver and shock/fire. To oppose it, they must concentrate in some fashion. Air power poses a similar problem from a different domain. To oppose it, they must disperse, hide, dig in, and so forth. If adversaries concentrate to oppose land power, air power will destroy them. If they disperse to defend against air power, land power will destroy them. *Together*, U.S. air and land power can be unstoppable.

Air power can be concentrated rapidly, deployed swiftly, and strike with intensity and high accuracy. Together with rapidly arriving ground forces, the United States possesses a combined arms capability that should deter, but will also defeat, potential adversaries. Air power and ground forces offer two complementary capabilities, two forms of maneuver and concentration that ought to be unstoppable.

The question is, how should the merits of air and space power be exploited in conjunction with other forces to provide the United States with unstoppable military power? Further, what must land forces do to take advantage of modern air and space power? One approach is faster arrival in theater. The combination of fast-arriving, fast-engaging aerospace and land power denies the adversary the advantages he will inevitably seek in time, space, concentration, dispersal, concealment, and the like. It will also limit the opponent's ability to successfully pursue asymmetric strategies. Together, air defenses, bombardment, and land operations denied the Germans the use of V-1 launch sites. In the future, the rapid and decisive employment of air and ground forces should deny a regional aggressor the ability to use ballistic missiles in a politically or militarily useful fashion.

In defining the merits of modern air and space power, the study's participants recognized that those merits must be appreciated at three levels. First, a revolution in military technology has the increased capability of air and space forces to see, maneuver, and strike targets with great rapidity and efficiency. This revolution both enhances the performance of air- and space-oriented capabilities and allows for a significantly enhanced exploitation of the vertical dimension of warfare, thereby moving from a concept of warfare based on the battlefield to one based on battlespace. Second, these new capabilities or performance improvements provide the means to dominate the aerospace domain and from there increasingly influence the course of events on the earth's surface. Third, and perhaps most important, the ability to dominate air and space and to operate continuously and fully from that position creates the opportunity to adopt a new, three-dimensional approach to warfare, that is, warfare whose base for operations is "located" in the third dimension. From what we are calling the vertical perspective, modern air and space power can enable the United States to dominate its adversaries across much of the conflict spectrum. The strategic implication of this new approach is that it permits the conduct of military operations in a manner consistent with our strategy and interests while denying an adversary a viable course of action save cessation of hostilities or defeat.

A New Paradigm for Twenty-first-Century Warfare

The Revolution in Military Affairs: An Air and Space Power Perspective

There has been a protracted debate in this country over the past several years about the prospects for what is termed a Revolution in Military Affairs (RMA). An RMA is a substantial change in the means and methods of warfare, such as that associated with the advent of gunpowder, the tank, the airplane, wireless communications, and nuclear weapons. Proponents of a modern RMA generally point to a set of technologies that can substitute for existing military systems with greater effectiveness and/or lower costs as evidence of the revolution. At present, most proponents believe that the explosion of information technologies and their increased role in both military and civilian systems indicate that we are, at the very least, on the threshold of an RMA. In the military sphere, stealth, precision guidance, directed energy, and multisource sensing are all cited as examples of an emerging revolution.

But a revolution is more than just the introduction of new technology. As RMA proponents point out, the true revolution occurs when new technology promotes the creation of new forces and their application in new ways. Many nations are capable of developing the technologies of war and even deploying them in the field. What is revolutionary is when those forces are organized and employed in ways that provide significant leaps in the effectiveness and power of military forces. Modern mechanized warfare, the Blitzkrieg, based on exploiting the capabilities of tanks, aircraft, and wireless communications, is often cited as an example. So, too, is the application of air power to naval warfare, beginning in World War II, which led to the creation of carrier battlegroups and the conduct of naval engagements at ranges previously considered unthinkable. Thus, revolution that transforms the conduct of warfare, the RMA, is not one of technologies or weapons systems, but one of perspective or viewpoint. Although it is enabled by technology, and information technologies in particular, this revolution has much more to do with how military forces perceive and address their environment.

The experience of naval doctrine and strategy since the advent of air power is illustrative of how new capabilities can create a change of perspective. When compared to land warfare, warfare at sea is similar to that from the air insofar as neither medium has physical boundaries or limits. Forces are not based at sea but are deployed to sea, where they exert influence over battlespace by virtue of the ability they have to see and act in that space. Historically, the role of naval power and limits to concepts of naval operations were determined by the distance that naval guns could fire. The purpose of naval maneuver was to enable separate units to concentrate fires. Although naval power could, to a very limited extent, affect the shore, its fundamental purpose was to establish control over the seas, sweep them clean of enemy forces and shipping, and secure control of the seas and approaches to enemy territory. With the advent of naval aviation, not only did the role of naval forces expand, but the concept of naval strategy and the relationship between land and sea changed dramatically, albeit over time. Naval forces could now affect the land in a manner heretofore thought impossible. Conversely, land forces, specifically land-based aviation, could pose a threat to naval forces far out to sea. As the war in the Pacific demonstrated, rather than freeing naval forces from a connection with the land, air power created a closer relationship, requiring the seizure of islands across the central and southwestern Pacific in order to establish airfields from which land-based aircraft could be projected forward across the seas. With the introduction of aircraft, the requirement for information in naval operations skyrocketed, as did the need for data links and data fusion. Increasingly, success in naval operations, whether against other naval forces or against the shore, meant the real-time coordination of distant units performing a variety of missions to achieve information superiority, air superiority, and precision engagement of the adversary.

Modern air and space power can potentially change the perception of future warfare even more than it did for naval power almost 60 years ago. For millennia, mankind has lived, fought, and died on the surface of the earth. For many decades, even the ability to move through the air and under the seas did not alter the two-dimensional perception of existence. This perception, however, is changing at an ever-increasing rate, as technology allows us to exploit the opportunities created by operating in and through air and space. Our historical, horizontal perspective is slowly giving way to a new, vertical perspective. As we grow comfortable with this change, we grow increasingly comfortable with seeing our reality in three, rather than two, dimensions.

This new perspective has become part of everyday life. We are increasingly accustomed to living a three-dimensional existence based on the ability to operate above the surface of the earth. Indeed, this new perspective has become so familiar that it hardly seems revolutionary anymore. Weather information is received from space, not from weather systems located over the horizon. Space-based navigation systems beam down coordinates that allow travelers to ignore roadside markers. Television antennas are oriented not towards the horizon, but toward space. Geographic information services, which combine imagery and geolocation data gathered from space to form hyperaccurate three-dimensional maps for private and commercial purposes, are a growing industry.

The same revolution of perspective on viewpoint has influenced warfare. As the Gulf War suggested and what has become palpably clear in the ensuing years, technology is liberating warfare from a "horizontal" perspective. From the beginning of time, armies and forces aligned themselves against one another, maneuvering over the surface for advantage, to seize favorable or valuable terrain. Control of the intervening ground was critical to the ability to defeat the adversary, and thus precondition for attaining victory.

Air power, too, in its earliest days, was viewed from this horizontal perspective. The impetus to exploit air (and later space) power—to expand warfare to the third dimension—was part of a larger effort to overcome the epitome of the horizontal perspective of linear warfare as practiced in World War I. Air power was viewed as a method to restore to the battlefield maneuver via indirect and remote fires. During World War I, these fires were delivered principally by artillery, and air power was, not coincidently, concerned primarily with its support.[14] Since World War I, artillery fire has been complemented by a variety of means: infiltration tactics in 1917–1918, Blitzkrieg (1939–1940), parachute landings, and strategic bombardment in World War II; and airmobile and strategic air operations in the post-World War II era. In all cases, the goal has been the same: to restore momentum across the ground between one's own forces and their war objectives. Over time, the success or failure of these efforts has been a function of the nature of the theater (its size and composition), the technical and quantitative character of the engaged forces, and the scale of the conflict. Often, success on a relatively small scale, such as in the German offensive of March 1918 and the campaign in France 1940, could not be sustained when the scale of the operation was expanded or the theater was too large.

Throughout this period, air power was a captive of horizontal perspectives, differing from surface combat only in that air combat was conducted at altitude. Despite the evident increase in the performance of air and space systems, our view today of its contributions remains largely dominated by a horizontal perspective. Technological limitations and inadequate control of the airspace compelled air forces to apply their resources en masse, in formation and in a series of battles and campaigns.

The limitations of air power, whether employed tactically or strategically, inevitably tied it to the war at the front lines. Here, by all accounts, air power often could be decisive, creating conditions that permitted the defeat of adversaries in detail.[15] By contrast, the effectiveness of air power applied to strategic bombardment, from World War II onward, is still subject to extensive debate.[16]

Experience in the Gulf War suggested the potential for achieving a new perspective. For example, the effort to counter Iraqi Scud missiles was conducted from a vertical perspective. It involved information on the detection of missile launches collected from space. This information had to be passed in real-time from Cheyenne Mountain, Colorado halfway around the world to command centers and missile batteries in Israel and Saudi Arabia. Ultimately, it was also fused with other data and transmitted to aircraft operating against Scud launchers in Iraq.

What this experience revealed was that the United States has the ability to perceive the world from a vertical perspective and communicate the knowledge so gained in a useful form, instantaneously, anywhere in the world. Moreover, it demonstrated that we can act on that information. The combination of new, advanced technologies (C^4ISR, navigation aids, airframes and power plants, stealth, smart weapons, new warheads, decision aids, modeling and simulation, and training systems) provides the basis for seeing and understanding the theater/battlefield in a new way and for acting on that understanding in near-real time. As these various technologies improve, air and space power, as its potential was glimpsed in the Persian Gulf, holds out the promise of responsiveness that may be the near-equal of continual presence.

This vertical perspective has allowed us to appreciate the potential of the third dimension—the domain of air and space power. This domain is more than a medium through which indirect fires can travel. Properly understood and exploited, it can become a base for operations. Freed from the conceptual limitations of a horizontal perspective, we realize that power can now be applied in all three dimensions simultaneously. This potential to transition from the serial or sequential warfare (even when making use of air and space power) that was characteristic of the horizontal perspective to parallel operations conducted simultaneously and continuously based on a vertical perspective, constitutes the essence of the RMA.

Defining a New Paradigm

The conduct of warfare from a base for operations "located" in the third dimension constitutes a new paradigm. This paradigm is made possible by utilizing a vertical perspective, which "sees" all three dimensions of warfare simultaneously. It depends on air and space power to provide the means to act decisively in all three dimensions in combination with land and sea forces. This paradigm would organize and apply emerging capabilities in mobility, information superiority, air superiority, long-range strike, and new logistics concepts to a three-dimensional strategy.

The combination of rapid seizure of air and space control, continuous real-time surveillance of the battlespace, and the conduct of around-the-clock precision operations poses a fundamental challenge to current concepts of theater campaigns with their ordered series of intermediate objectives. During the Gulf War, Allied air power had established the first base of operations in the third dimension and was conducting fully three-dimensional warfare. The simultaneous and continuous strikes against Iraqi ground forces, command and control assets, logistical networks, and infrastructure represented a break from classical concepts of air power doctrine. These concepts divided air operations into strategic attack, battlefield interdiction, and close air support. All three types of operations were conducted almost from the outset of the Gulf War. The weight of emphasis in the air assault was shifted over time based not only on the effectiveness of prior operations in destroying the intended target sets but also on evidence of the reaction of the enemy derived from continuous surveillance and intelligence assessment.[17]

A base of operations in the third dimension does not mean that assets are permanently deployed there in a fixed position. Rather, it is a virtual base, similar in concept to a naval theater, to which mobile assets are deployed and in which they operate over a period of time. Air and space assets would be deployed into the base of operations to perform missions. All of the missions associated with air and space operations—i.e., surveillance, intelligence and targeting, the seizure and maintenance of air and space superiority, the conduct of strikes at various ranges, the movement of forces and capabilities, and the support of operations—would be performed on an ongoing basis. How much force and of what kind would be present at any particular moment in the campaign would be a function of the overall friendly force posture in theater, the particular strategy being pursued, the adversary's actions and reactions, and the operations of ground and naval forces.

In the past, air power has been employed from a horizontal perspective to attack targets at the front or in the rear in order to influence action at the front. What is distinct about a base of operations in the third dimension is its ability to free joint forces from the requirement to engage in a contest to seize or control ground as the central means to achieving military and political objectives. Access to the entire theater is available from a position in the third dimension. Moreover, coordination of joint force operations is dominated by the ability to observe and act without the limits and hindrances imposed by geography and adversary forces on forces operating on a vertical perspective. As C^4ISR capabilities improve, so to will the ability to assess the impact across the theater of strikes against specific target sets and combinations of targets. Three-dimensional operations would coordinate the activities of fully joint forces to create opportunities for decisive action by a given arm or service, depending on the character and evolution of the battlespace. Thus, offensive ground operations might be initiated to force the concentration of adversary forces in a manner making them more suitable for destruction by indirect fire means. Similarly, the prior interdiction of field forces through the use of air power can create advantageous conditions for the initiation of offensive ground operations.

Looking at the theater battlespace in all of its dimensions from above, will allow theater commanders to pick places to deploy forces and power best calculated to achieve the objectives of the war. Developing a base of operations in the third dimension means concentrating combat power across space and time in a manner designed to defeat an adversary's efforts to employ his own forces and other assets and pursue one's own strategy. The generation of simultaneous operations from the third dimension across the entire theater against all target sets and the ability to sustain an extremely high tempo of air and space operations can stun, disorganize, and destroy the adversary in as much detail as is necessary to achieve friendly military and political objectives. The creation of a base of operations over the contested theater—involving the seizure of air and space superiority, the establishment of information superiority, the movement of forces into theater under the umbrella of air superiority, and the conduct of large-scale precision strikes—will almost without question constitute a de facto defeat of the adversary's strategy. If the initial effects of such three-dimensional operations have sufficient impact, producing a condition of "shock and awe," then further operations may be unnecessary.[18] If not, from the base of operations in the third dimension, the U.S. forces can undertake a campaign to progressively destroy the adversary's capability to employ military power.

This ability to dominate the third dimension, thereby overcoming the inherent limits of horizontal warfare, allows air and space power to be used as a force for operational and strategic maneuver rather than for attrition. From the third dimension, air and space forces deployed in-theater are inherently in the adversary's rear, astride his lines of communications, and surrounding his forces. Long-range systems provide strategic and both inter- and intra-theater maneuver as well. The capacity to conduct operational and strategic maneuver from the third dimension is a function, in part, of the ability of modern air and space systems to move faster, farther, and with greater survivability than was possible in the past. It is also true that technological advances in air and space systems have been such as to significantly increase the lethality of air power. Yet, although air and space power can visit devastating firepower against targets on land and sea, its ability to deploy force and assist in the deployment of non-air and non-space forces should be seen as one of its most significant values to the theater commander. The capability of modern air and space power to efficiently and swiftly employ force throughout the entire battlespace—a capability resulting from the coupling of information superiority and precision strike—can enable air and space power to be used as a maneuver force instead of an attrition force. The ability to employ air and space power as a maneuver force would be a complement to, not a substitute for, the operation of other maneuver elements that are part of a joint forces campaign.

The proposed new paradigm meets the challenge posed for U.S. strategy by its position as the reacting party having to cede the initial move to the adversary. Confronting a crisis or aggression, the United States seeks to rapidly engage the hostile party in order to immediately alter the situation in the theater or zone of conflict. This paradigm can be expected to increase the speed of U.S. responses, reduce expected U.S. vulnerabilities, and significantly enhance the overall combat effectiveness of U.S. military forces by setting our advantages against the opponent's weaknesses. Effecting change in the battlespace means at a minimum preventing an adversary from gaining an advantage from deliberately prepared offensive or defensive postures, such as those the Iraqis assumed immediately following the invasion of Kuwait. This approach would thereby seek to avoid the "set-piece" nature of the existing predictable model, in which the United States gains a lodgment and then attacks out of it, thereby giving the enemy time to establish his defenses and employ long-range strike systems before the United States can assemble an effective defense.

To rapidly impose force against an adversary and achieve this kind of change in theater, U.S. forces cannot wait for the opening of overt hostilities to begin operations that shape the battlespace and seize the initiative. Operational measures that are essential to gaining dominance of the conflict must be initiated before a crisis transitions to conflict. The new paradigm must recognize the need for pre-conflict activities to establish the basis for battlespace dominance and the wherewithal to conduct three-dimensional warfare. Information warfare, in addition to the technological advances noted above, provides a powerful set of tools to assist in these pre-conflict tasks, especially in light of concern over and potential constraints on preemptive actions. Air and spaces forces, because of their rapid responsiveness and global reach, will play critical roles across the entire C^4ISR and information warfare domains.

The initial application of combat in the new paradigm is designed to rapidly negate the adversary's strategy while simultaneously maintaining continuous pressure to force his *reaction* to our plan and tempo of operations. Once the adversary's basic plan has been thwarted and he is forced into a reactive mode, the initiative has passed into our hands. Thus, the initial operational moves must "set the terms" for the entire campaign.

Against adversaries deploying significant military power, the new paradigm would place early emphasis (from the opening of hostilities) on an integrated effort to win at least air dominance and fracture the cohesion of the opponent's forces. These are essential missions that must be accomplished before major surge and sustainment forces are placed in the immediate theater of operations and at risk. In addition, dominance of the battlespace will probably require destruction of the opponent's essential C^4ISR capabilities and substantial degradation of his critical defensive aerospace capabilities. These operations are designed to force the opponent into a "local" mode by disrupting the critical command linkages that allow the leadership to exercise "global" coordination. Even in the absence of a decisive advantage overall, the United States—enabled by its superior knowledge-based concepts of operation—could respond inside the enemy's decision and action cycle, thereby producing overwhelming local superiority time and time again. These actions, when combined, are designed to give U.S. forces the initiative in the timing and tempo of subsequent operations by putting the opponent in a reactive mode, continually pressing him, and not allowing him to regain command or the initiative in combat; they will help us achieve "battlespace dominance" throughout all mediums of combat operations. The United States must have the capacity to do this against future opponents that are more agile and effective than Iraq during Operation Desert Storm.

Once U.S. forces have seized the initiative and rendered the adversary's war plan ineffective, they can exploit these advantages, as necessary, to further the defeat of the adversary's military capabilities. Battlespace dominance, the ability to exploit the third dimension and operate across the entire theater, provides U.S. forces with the opportunity to fragment enemy forces. Such a fragmentation operation will allow the United States to defeat the opponent "in detail," without having to face concentrated forces or fires or attempting to force a single "decisive" action. Under these conditions, old images of linear combat and massed formations need to be replaced with an understanding of the dynamics of widely dispersed combat operations conducted in a sparsely populated battlespace by fleeting targets.

This new paradigm depends on a fully integrated operation with advanced C^{4}ISR capabilities as its backbone; and it requires close coordination and synchronization among all operating elements: sharing a common tactical objective, employing common doctrine, operating in close cooperation with tactical echelons, and providing mutual support. Diverse sets of force capabilities (such as air and ground) can be directed to accomplish a single tactical objective and are coordinated in space and time to achieve a decisive tactical result that can be exploited in this nonlinear environment to produce significant operational victories. This integration enables sustaining a high tempo of operations, which, when combined with responsive adaptive forces, will allow U.S. forces to anticipate the enemy's decision and action cycles. These real-time mutually supporting adaptive capabilities will create unexpected opportunities because the opponent will never quite be in step. The opportunity to initiate operations with a minimum of deliberate planning or planning or preparation means that U.S. forces can exploit opportunities created in this complex battlespace. In this concept, as opposed to the more traditional views of air and space power as the supporting force, air and space power can be the principal lethal element in some future conflicts. Further, this concept would maximize the ability of air power to be devastatingly lethal against massed forces by combining them with ground maneuver elements that compel the opponent to concentrate or be defeated in detail by the integrated force.

The concept of three-dimensional warfare is not limited to the operation of air forces alone. Each of the armed services is increasingly exploiting the advantages of the third dimension to better conduct their own operations. No two services will be equally invested in air and space power, yet each will be critically dependent on the advantages proved by the occupation of the air and space domains and by the establishment of continuing operational capabilities in the third dimension. Rather than merging military forces, modern air and space power provides new means for each service to pursue its own form of warfare but with greater effect and efficiency. The U.S. Army and U.S. Navy are both moving critical operations into the third dimension as a means of empowering their horizontal operations. The Advanced Warfare Experiment (AWE) associated with the army's Force XXI initiative demonstrated the impact for decisive maneuver of information superiority, UAVs, and heli-borne forces. Similarly, the navy is developing the cooperative engagement capability (CEC) to leverage its existing capabilities in C^4ISR.

Aerospace operations (for both air and space forces) would move from a preplanned episodic/periodic model to one that stresses real-time adaptivity, continuity, and persistence. The new operational patterns that these would allow would provide a significant opportunity to develop a fully integrated joint operational approach that maximizes the potency of U.S. forces and enables politically effective operational results without overwhelming force. Continuous air and space operations, instead of episodic appearances in the battlespace, would allow air power to create an "air overwatch" capability to dominate all tactical operations beneath. One option is full integration through an expanded, enhanced CEC. These new concepts would utilize many of the same technologies demonstrated during the Gulf War—stealth, effective SEAD, proliferated space sensors, real-time airborne battle management—but they would be integrated into an operational concept for continuous air operations in order to dominate all actions in the battlespace.

The Strategic Implications of the New Paradigm

Modern air and space power employed in the service of fully three-dimensional warfare should allow us to dominate, control, and shape any theater of war, from Haiti, to the Kuwait theater of operations/Iraq, to that which would exist if we confronted a peer competitor. The United States seeks to establish control at a theater level regardless of the character of the conflict. In most cases, theater-level control involves pursuit of a number of objectives by a variety of means. It can involve sanctions enforcement, sea and air space control, movement interdiction, information operations (both offensive and defensive), strikes against a wide range of targets, and combined-arms, land-oriented battles.

The use of air and space power in three-dimensional warfare is central, even singular, to the establishment of theater-level control. Establishing theater-level control means being able to determine what the adversary can and cannot do within the theater. This control can take a variety of forms, but control of the air space above the theater and, by air power, over what happens on land and sea is the most important and potentially decisive form. Air power is the force most capable of addressing the range of goals and missions associated with controlling a theater. As future adversaries become more fully capable of exploiting the attributes of long-range indirect fire systems, thereby increasing the scale and complexity of future theaters of war, space power will increasingly become a critical element in the attainment of U.S. air and space superiority.

The ability to control the theater also provides the National Command Authority (NCA) with the ability to respond to unforeseen and even very difficult crises. For example, if North Korea presented a missile threat to Japan, air and space power would provide the NCA with a range of responses up to and including the invasion of North Korea. Absent air and space power, only an invasion would be able to stop the missile attacks, and such an action would be problematic at best.

The ability to control a theater is not the same as the ability to shape it. Air power offers the greatest opportunity and flexibility for shaping a theater. It does so first by the threat of action—to seize air superiority over the theater or to destroy adversary ground and sea units. Second, it constrains the range of actions the adversary can take in the face of air power. Third, it enables other elements of the friendly/U.S. force posture to act not merely as a force multiplier but as a necessary or even vital independent element in their operations. For example, the presence of friendly air means that ground forces can operate more effectively. Conversely, adversary ground forces that disperse to meet the air threat are more vulnerable to ground forces, and those that retreat or concentrate in the face of a ground offensive are more vulnerable to air power.

The idea of control at the theater level does not lend itself to the halt, buildup, counteroffensive paradigm that is the current fashion. Nor should it. The choice of approach is up to the commander in chief (CINC). Logically, the way he will fight the battle and the length of time for its phases (should they occur) will be determined by the ability to control, shape, and dominate the theater. If hostile forces are attrited, if they retreat unexpectedly, or if they can be rendered harmless by indirect means, then the need for the complex buildup and counteroffensive is considerably reduced. This does not mean, however, that air power at the conceptual level needs to conform to the three-part planning paradigm.

Although the discussion to this point has focused on the merits of air and space power in war, they are, in character if not in the same exact form, applicable to other types of operations. Clearly, the capacity to operate in three dimensions, to see the battlespace or the theater in its totality, to have the long-range reach provided by global strike and mobility forces and the ability to achieve focused effects without mass, can contribute in numerous ways to the full range of lesser operations: humanitarian relief, operations-other-than-war, military operations in built-up areas, and so forth. In addition, modern air and space capabilities can now enable new types of operations, such as those over the former Yugoslavia and Iraq, that are intended to deny aggressors the ability to use airspace in pursuit of their offensive aims. These denial operations constitute a peacetime air equivalent of wartime sea blockades.

From a strategic-political perspective—one that considers the goals of war—the new paradigm offers the prospect for a new doctrinal approach to warfare, one focused on attainment of one's objectives rather than on denial of the adversary's objectives. Denial operations may have to be conducted during the course of a conflict. Nevertheless, it is the ability to achieve ones own objectives and to sustain a position of control over the adversary that will determine success at the conclusion of hostilities. Damage infliction is a means to attaining one's operational and strategic objectives but, increasingly, is likely to be a second-order means.

Earlier efforts to use air power as a means of achieving political objectives in war have centered on air power's coercive capability measured in terms of destructive power. Some have argued that modern air and space power alone make it possible to successfully pursue political objectives by threatening to attack or by actually destroying an adversary's most important and most valued assets.[19] This may now be possible in some instances. Emphasis on the purely coercive role of air and space power runs directly counter to its merits, however, and to the potential offered by the new paradigm of three-dimensional warfare. Moreover, campaign planning based on the idea of coercion through pain infliction may undermine one's ability to employ modern air and space power in ways that effectively exploit its capabilities to control the battlespace, shape adversary behavior, and bring the conflict to a more rapid conclusion. The measure of effectiveness for a future joint campaign centered around air and space power is the attainment of goals irrespective of decisions made by the adversary. Efforts intended to affect the adverary's will, as opposed to his capacity to act in opposition, are more likely to prolong hostilities than to elicit compliance.

Focusing on the capacity of air and space power to coerce through the infliction of damage as evidence of how well it supports the attainment of national objectives reverts, in essence, to a two-dimensional view of war, with aircraft taking on the role artillery played in World War I; that is, to conquer so that the infantry could occupy. As we discovered through twentieth-century warfare, the choice of acquiescence or capitulation is ultimately in the adversary's hands. The opponent can choose to endure artillery or air bombardment, not only foiling the other side's strategy but also, in effect, seizing the initiative: the adversary is likely to know his own pain threshold considerably better than his attacker does. "Bombardment has an ambivalent nature," wrote one of the founders of modern air power theory. "At certain times and places it may produce positive and at others negative morale reactions."[20] The historical record on the power of bombardment alone is not optimistic. In World War II Leningrad withstood 900 days of siege and bombardment without capitulating; Berlin surrendered only under direct Soviet assault; and Japan steadfastly refused to surrender despite the effects of both strategic bombardment and sea blockade. Witness also the resistance of the Chinese and North Koreans during the last stages of the Korean War; the years-long campaign against North Vietnam; and the apparent lack of response by Saddam Hussein to the threat and reality of massive air assault.

The power of three-dimensional warfare is that it does not rely on coercion of an adversary to achieve strategic objectives. Instead, it provides a means for imposing control over the adversary's behavior, as distinct from his intentions or will. The imposition of control can, over time, progressively shape the adversary's actions, deny response options to one's own actions, and deny the adversary the capacity to conduct effective military operations. Three-dimensional warfare offers an endless combination of operations against an enemy that are independent of his choice to continue hostilities. Although those operations are aimed at forcing the adversary to conform to one's strategic and political objectives, they do not depend on his decision to cease hostilities. In that sense, it is not dependent on breaking his will. The combination of operations possible with three-dimensional warfare permits one's own forces to achieve objectives valuable to them even as it denies assets of value to the adversary. In the end, it should not matter whether the adversary chooses to capitulate, as he is rendered progressively inert and helpless. This alone may be sufficient to "break his will." Ultimately, however, the adversary's determination to continue his aggression or resist his opponent's will is irrelevant in light of a properly constructed war plan that effectively exploits modern air and space power.

The possibility of exploiting the air and space domain—the third dimension—as the base for military operations to create a new way of warfare has not been realized until now because of technological limits and the inability to meld its components (C^4ISR, maneuverable and on-call indirect fires, and decisive maneuvers in three dimensions) successfully. Today, the United States is over the threshold on technology and about to cross it with respect to melding the requisite components to achieve a decisive strategic advantage over potential adversaries. We are standing on the verge of a revolution in warfare whose promise ought to shape the pending restructuring of the nation's armed forces.

Realizing the Revolution

The dramatic change that has taken place in air and space technology over the past few decades offers the prospect for a revolutionary shift in our understanding of air and space power and how it can be employed in warfare. Technology has empowered air and space forces to perform traditional missions with greater efficiency and effectiveness. But modern air and space power is more than just the aggregation of improved technical capabilities. It involves a way of thinking about the use of military force across a three-dimensional battlespace.

The U.S. Armed Forces have been moving at a pace set by the evolution of technology, budgetary constraints, and the adaptive capacity of large institutions, toward adopting concepts derived from a vertical perspective. One of the earliest efforts in this direction was the AirLand Battle doctrine. It focused on the use of advanced sensors and air and missile systems to create gaps in the adversary's echelonment of forces, which in turn could be exploited through a combination of maneuver and fires.[21] Another was the U.S. Navy's Maritime Strategy, which relied on overhead assets, maneuvering naval forces, cruise missiles, and manned aircraft to exploit weaknesses on the Soviet Union's flanks in the event of war. The report of the Commission on Integrated Long-Range Strategy, *Discriminate Deterrence*, represented yet a third approach. This report proposed a U.S. security strategy that emphasized flexibility in a changing world while addressing critical threats to the nation. A central concept was the use of emerging capabilities in surveillance and targeting in combination with long-range offensive assets to deter and defeat regional adversaries.[22]

In the last few years, the Department of Defense and virtually all of the armed services have begun to experiment aggressively with new concepts for three-dimensional operations. Centered on the exploitation of advantages in the third dimension, each concept relies on control of the air and space domains and the ability to operate in and through these domains to provide intelligence, targeting, remote fire support, and logistics to forward deployed forces.

This recognition of the potential for a revolution in warfare is validated by *Joint Vision 2010*. Central to the concepts of this vision is the ability to exploit the third dimension to support all military operations. The goal is to achieve "full spectrum dominance" by taking advantage of our capabilities in remote fires, information superiority, mobility, and maneuver that are, in essence, synonymous with the characteristics of modern air and space power.

The U.S. Marine Corps is pursuing the Sea Dragon experimental concept, an effort to exploit advances in remote sensing, unmanned vehicles, long-range on-call fire support, lean logistics, data fusion, and long-distance communications to allow forward deployed units to multiply their mobility and lethality. The U.S. Army is focusing on Force XXI and then the Army After Next, both of which may employ some of the same concepts under evaluation by the marine corps. It is an effort to enhance the utility of traditional platforms, such as tanks, self-propelled artillery, and helicopters, by providing them with real-time digital data links, navigation aids, and computers. The U.S. Navy is focused on the idea of expeditionary power projection and the use of advanced technologies to dominate not only the seas but also the littorals. Concepts for a so-called arsenal ship, as now seen in the future surface combatant or SC-21, rely on third-dimensional capabilities for sensing and shooting. The navy has also led the way in the integration of sensors and weapons in its cooperative engagement capability. Field tests of this concept have demonstrated our ability to use sensor data from one source to employ weapons coming from another platform.

The U.S. Air Force has also grappled with the implications of the revolution in air and space power. Most recently, in its *Global Engagement: Vision for the 21st Century*, the air force demonstrates a commitment to operate in the third dimension. *Global Engagement* was shaped by assertions that the air force is in a transition from an "air force" to an "air and space force" and then on an evolutionary path to a "space and air force." It acknowledges the migration to space of several key functions: intelligence, surveillance, and reconnaissance (ISR), warning, position location, weapons guidance, communication, and environmental monitoring. Even at this relatively early stage in its metamorphosis, the growing capability of air power to provide global situational awareness, unparalleled reach, swift and deadly precision strikes, and massive mobility and logistics for power projection creates a new type of capability for joint forces commanders. *Global Vision* recognizes that it will be possible for joint forces applying modern air and space power to find, fix, track, target, and destroy targets on the land, at sea, and in the air. Applied strategically, this capability provides the basis for conducting theater-level strategic campaigns to defeat an adversary's strategies; shatter the cohesion of its forces and support systems; destroy critical military, command and control and industrial systems and establish the basis for dominance at the strategic level. The ability to bring intense firepower to bear over global distances within hours to days gives national leaders unprecedented leverage.

Although these doctrinal and technological experiments show promise, they have not yet been fully freed from the horizontal perspective. For the army, marine corps, and the navy, the cognitive base of operations is the surface of the earth. Their use of the third dimension is intended to overcome the obstacles erected by an adversary to impede their progress over the surface. Air power is still viewed as "indirect fire"; space power is viewed as support to and enhancement of capabilities whose primary orientation is to control the area between one's own forces and the objectives of the war.

Nevertheless, the promise of revolutionary gains made possible by a shift in perspective has continued to lure analysts toward adopting a truly vertical perspective. For example, in the Office of the Secretary of Defense they are struggling with the potentially revolutionary implications of a vertical perspective on conflict. The 1996 Defense Science Board summer study, *Tactics and Technology for 21st Century Military Superiority*, identified concepts for the exploitation of emerging capabilities to free ground forces from a fixed base of operations by providing them with on-call (nonorganic) long-range indirect fires and a robust information network based on airspace communications to increase substantially the power of early deploying forces.[23] It is reasonable to suppose that many of these same emergent capabilities explored for small units in low-to-medium intensity engagements can be employed in larger or more intense combat situations.

Douglas A. Macgregor's recent book, *Breaking the Phalanx*, gives this supposition substance. This new concept for the army of the future proposes to exploit the advantages offered by control of air and space by organizing units around flexible, mobile "combat groups." From the vantage point of a vertical perspective, both remote fires and enhanced direct fires would be used to create dominant maneuver. The author envisions the close coordination of air power and more mobile ground forces, including substantial airmobile ground forces, to conduct early entry as well as dominant maneuver operations. Control of the air and space domain would also provide the means to create powerful air assault forces—veritable vertical maneuver groups—that could conduct three-dimensional maneuvers.[24]

Realizing an RMA based on a full explanation of the merits of modern air and space power involves three basic "steps." First, it means investing in critical capabilities such as appropriate C^{4}ISR architectures and systems to enable all forces to make use of a vertical perspective and the information provided from the vantage point above the battlespace to plan, control, and direct forces. Other critical capabilities include the platforms and weapons systems that will ensure dominance of the air and space domains. Air and space domination will require systems that are capable of striking targets in extremely high-threat environments from the outset of hostilities. Exploitation of the inherent attributes of global attack and precision strike will inevitably require weapons that have greater precision targeting capabilities and can be launched from standoff ranges of increasing distance from their targets.

The Working Group on Technology and the Industrial Base identified a number of critical technologies that need to be more fully exploited or where the current U.S. lead must be maintained. Among these technologies are survivability, air vehicle stealth, mobility assets, space-based survivability, all-weather strike (both targeting and PGMs), and noncooperative Identification Friend or Foe (IFF). Distressingly, these critical technologies by and large are of limited interest to the civilian market. Nevertheless, they should receive both political and financial support. Their export should be undertaken with extreme caution, and a continuous review of both prime and lower-tier suppliers should be made to ensure that a healthy competitive market for such technology remains.

The second step is to change the way forces are employed. The proposed new paradigm focuses on the centrality of operations that exploit air and space power to seize the initiative from an adversary, foreclose his strategic options, and progressively destroy his capacity to wage war. The domination of the third dimension and the ability to operate throughout the battlespace and in all three dimensions also means the abandonment of the horizontal approach to organizing and deploying forces. Information superiority, precision strikes, mobility, and theaterwide maneuver provide a basis for conducting a campaign of focused attacks intended not simply to destroy valuable targets but to "disassemble" the adversary's war-waging capability as well. The coordination of force employment based on information, movement, and strike should enable forces exploiting air and space power to use the adversary's own power and actions against him. Maneuver by friendly forces can elicit responsive motion from the adversary—that is, movement and concentration of his forces—thereby creating opportunities for their destruction.

In combination, ground, sea, and air forces can be used strategically to establish new conditions on the ground, for example, a second front in the enemy's rear. To illustrate, preemptive occupation of Euphrates River crossings would have fundamentally changed the course of the Gulf War. With air and space superiority, mobility, and the ability to employ long-range indirect fires, a superior air and space force can pick targets not only for air strikes, but also for penetration operations. The implications of the freedom to act in three dimensions is the ability to create a "three-dimensional mobile group." Such an approach would probably require subordinating some air and space power assets to the mobile force commander while employing other assets on a theaterwide basis to create the conditions that would support operations beyond traditional front lines.

The idea of conducting integrated, continuous operations using the full range of available air and space assets raises the question of who should control joint air and space forces. To provide the greatest opportunity for achieving the scale, scope, and tempo of operations envisioned by the new paradigm, the theater commander probably will require a single joint force commander responsible for the employment of air and space forces, at least early in a campaign. Establishing the proper scope, scale, and tempo of operations, seizing the initiative from the adversary, and ensuring the desired progression and outcome of the conflict is likely to require centralized management of air and space power. As the campaign progresses, the importance of a joint aerospace component commander relative to that of ground forces or naval forces commanders may change.

The development of new types of operations and new command relationships also will require a program of games and exercises. To be effective, these games and exercises will have to be conducted jointly, preferably as theater-level campaigns. Noticeably absent from the above descriptions of DOD efforts to define future forces and operations are new concepts (and the follow-on games and exercises) on a joint basis. The problem with building modern air and space power from the bottom up is the difficulty it creates for architectural and operational integration at the theater or campaign levels. For this reason, top-down coordination of the various DOD and service-led efforts to exploit the revolutionary potential of modern air and space power is deemed highly desirable.

The third step in realizing the revolution is to redesign the forces in light of the new technology's potential and the evolving nature of operations designed to leverage air and space power. Intra-theater mobility and inter-theater maneuver will become more critical. Exploitation of the advantages in the battlespace offered by modern air and space power is likely to call for the development of forces that can respond to the faster tempos and greater scales of future campaigns. Forces will need to be lighter, have fewer logistics requirements, possess greater agility, and wield more firepower per unit of measurement.

It seems clear that joint forces doctrine and future theater campaign plans will need to adapt in order to take advantage of new tactical and operational opportunities created by the use of air and space power. The paradigm for future warfare proposed herein is inherently joint. The exploitation of the integrative perspective available through information superiority and full use of the third dimension must act as a magnet for joint action at the campaign or theater level. At the same time, all of the services will be able to use operations and assets in the third dimension to increase the effectiveness of their own forces. Circumstances at the time operations commence will still determine the relationship between forces. The operations of other forces will be enhanced and the overall campaign conducted more effectively if other forces organize themselves according to what air and space power can do. This potential of modern air and space power is understood by military organizations around the world and drives their efforts to reshape themselves for twenty-first-century warfare.

Space control and operations in and from space will play an increasingly prominent role for a military that seeks to exploit the potential of modern air and space power. The U.S. Air Force's long-range plan envisions the evolution of that institution from an air and space to a space and air force. The growing importance of space-based assets to air and space superiority means that those assets will become targets of hostile action, requiring that they be defended. The logic of the new paradigm suggests that there will also be a future need to deploy weapons in space that are capable of taking advantage of information superiority to more rapidly secure air dominance and to increase the tempo of precision strikes. This, in turn, means changes in how we organize to operate in space.

There inevitably arises the question of what is the appropriate acquisition strategy to support a transformation of the magnitude and character called for by this essay. The Working Group on Technology and the Industrial Base underscored the need for a comprehensive and well-funded program to ensure that critical technologies are brought forward and fielded. The panel expressed concern that the underfunding of basic research and development (R&D), the 6.1–6.3 programs, threatened to limit the ability to translate promising, innovative ideas into real systems. In addition, the probable scarcity of procurement funds in an era of balanced budgets called into question DOD's ability to translate technological promise into actual fielded capabilities. This problem is particularly worrisome in light of the growing availability to potential adversaries of dual-use technologies with significant military applications.

The development of joint and service-specific new military doctrines reflecting the changing character and new merits of modern air and space power will require new measures of effectiveness (MOEs) and modeling and assessment methods. Traditionally, MOEs for air and space power have been focused on maximization of the technical performance of the system (e.g., sustained rates of fire, maximum speed, operational ceiling), or force maintenance (e.g., sortie rates, ton/miles) or attrition-oriented force-on-force impacts (e.g., enemy aircraft destroyed, targets struck, tonnages of ordinance delivered). It has been difficult to translate these MOEs, even when adequate and accurate data has been available, into operational or strategic gains. The operational and strategic attributes of modern air and space power as conceptualized in the proposed new paradigm—the capacity to defeat an adversary's strategy (as opposed to his forces), to establish control over the battlespace, to shape the adversary's behavior throughout the course of the conflict and to impose one's will—are incomparably more difficult measures of effectiveness to define and assess. Conceptually, a more appropriate MOE for the new paradigm would relate operations and missions achieved by one's own forces to operations denied to the adversary; the attainment of a satisfactory level for one or the other, or a combination of the two, leaves the side pursuing this new doctrine satisfied that it has achieved a successful outcome of the war. An appreciation of the merits of modern air and space power necessitates an effort to experiment with the new paradigm proposed herein and to develop the analytic concepts, tools, and techniques to support development of new military doctrines.

Conclusion

We return to the question that began this discussion: should air and space power be made more central to strategic and operational planning? On the basis of the assessment provided in this study, the answer is yes. Increasingly, the capabilities of modern air and space forces make the conduct of military operations without their employment virtually unthinkable, even impossible. The kinds of air and space capabilities that are being deployed or that can be created offer the prospect that future air and space operations will be conducted in a manner, at a pace, and with an efficiency hitherto unknown. Moreover, these capabilities offer a greater flexibility and freedom of action in a new and more difficult emerging strategic environment.

More important, air and space power can serve as the basis for creating a new approach to war. As argued in this study, air and space power has reached a state of technological and operational maturation that enables it to form the basic building block of a new strategic paradigm. That paradigm would see warfare conducted increasingly from a base for operations located in the third dimension. All forces would use that base to pursue their specific objectives, but not all operations would be located in the third dimension. Three-dimensional warfare presents numerous opportunities to affect an adversary's will and achieve one's own objectives. Overall, the result would be a national military capability that is comprehensive in its character, global in its reach, swift in its response, and highly effective in its actions.

This new strategic paradigm will require the investment of resources, as well as time, to fully develop the capability to see from a vertical perspective and operate in three dimensions. It will depend on our ability to leverage the progress already made in exploiting air and space and to extend those successes. Further, it will depend on the deployment of new capabilities, some of which are only beginning to be available and others of which are not even fully developed in the laboratory.

This study has reconfirmed the fact that air forces are increasingly essential instruments of military power. This is particularly true of American air forces today. In the future, as they evolve with new and more capable systems entering the force, they will become even more important. Their ability to accomplish a growing set of missions with an ever-increasing degree of certainty, less vulnerability to themselves, and reduced collateral damage is unquestioned in this essay and should be unquestionable based on the events of recent years.

The historical record and recent experience both tell us that air and space power has been most effective when applied en masse in conjunction with other forces. Nevertheless, there remain the questions of whether air forces can be the principal element of a future theater campaign plan, and, whether they can offer the prospect for achieving the campaign's objectives largely through air and space operations. The historical record, suggests that air forces have not been indisputably successful in the strategic role long sought by air power advocates. The debate between air power advocates—who have asserted that air power has never been given the time, resources, or freedom of action to fully exploit its inherent power—and the skeptics—who argue that air power by itself or in a leading role is neither sufficiently lethal nor effective to break the adversary's will or defeat his military forces—has yet to be resolved.

This study takes the view that the effort to resolve this debate misses the real potential that air and space power may have to recast the way in which war is waged. The undisputable capabilities residing in modern air and space power can, by overcoming the impediments and obstacles of conflict (time, distance, mass, and visibility), serve as the basis for defining the future character of all of the U.S. armed forces.

Notes

1. Stephen A. Cambone, "The RMA and Continental-Size Adversaries: A Case Study" (unpublished manuscript, January 1995).

2. Russell Weigley, *History of the United States Army* (New York: Macmillan, 1967), 411–414. See also Leonard Baker and B. F. Cooling, "Development and Lessons before World War II," in *Case Studies in the Achievement of Air Superiority*, Benjamin F. Cooling, Center for Air Force History (Washington, D.C.: GPO, 1994), 1–64.

3. Allan Millet and Williamson Murray, *Military Effectiveness, Volume II: The Interwar Period* (Boston: Unwin Hyman, 1988).

4. John McCarthy, "Air Power as History: Looking Backwards to Looking Forward," in *Smaller But Larger: Conventional Air Power in the 21st Century*, ed. Alan Stephens (Canberra: P.J. Grills, 1991), 23–29. See also Ronald Spector, "The Military Effectiveness of the U.S. Armed Forces, 1919–1939" in *Military Effectiveness*, ed. Millet and Murray, 83–84.

5. *The United States Strategic Bombing Survey, Overall Report (European War)*, September 30, 1945, 1 (hereafter cited as *USSBS*).

6. Ibid., table 1, p. x.

7. Ibid., Conclusions, 107–108. In its commentaries the *USSBS* noted the limits of strategic bombardment in terms often employed by modern critics: the resilience of the adversary, the limited effects of bombardment on morale, the inadequacy of strategic intelligence to support effective targeting, and the costs associated with maintaining control of the air. See also Allan R. Millet, "The United States Armed Forces in the Second World War," in *Military Effectiveness, Volume III: The Second World War*, Millet and Williamson Murray (Boston: Unwin Hyman, 1990), 61–62. For a positive case regarding the role of strategic bombardment, see also Richard Overy, *Why the Allies Won* (New York: Norton, 1995), chapter 4, "The Means to Victory: Bombers and Bombing." See also Robert A. Pape, *Bombing To Win: Air Power and Coercion in War* (Ithaca: Cornell University Press, 1996).

8. Mark Clodfelter, *The Limits of Air Power: The American Bombing of North Vietnam* (Princeton: Princeton University Press, 1989). See also Wayne Thompson and Bernard C. Nalty, *Within Limits: The U.S. Air Force and the Korean War*, Air Force History and Museum Programs (Washington, D.C.: GPO, 1996), 44–57, and John Schlight, *A War Too Long: The History of the USAF in Southeast Asia*, Air Force History and Museum Programs (Washington, D.C.: GPO, 1996), 45–43, 97–100.

9. See, e.g., the series of essays in Richard H. Schultze Jr. and Robert L. Pfaltzgraff Jr., eds., *The Future of Air Power in the Aftermath of the Gulf War* (Maxwell AFB: Air University Press, 1992); Richard G. Davis, *Decisive Force: Strategic Bombing in the Gulf War*, Air Force History and Museums Program (Washington, D.C.: GPO, 1996); Col. David A. Deptula, *Firing For Effect: Change in the Nature of Warfare* (Arlington, Va.: Aerospace Education Foundation, 1995); and Elliot A. Cohen, "The Meaning and the Future of Airpower," *Orbis* (Spring 1995): 189–200.

10. Elliot A. Cohen, *Gulf War Air Power Survey: Summary Report* (Washington, D.C.: GPO, 1993).

11. Information superiority is defined by the Department of Defense as "the capability to collect, process, and disseminate an uninterrupted flow of information while exploiting or denying an adversary's ability to do the same." Department of Defense, DODD 3600.1, *Information Operations*, December 9, 1996.

12. Richard P. Hallion, "Precision Guided Munitions and the New Era of Warfare," *Air Power History* 43, no. 3 (Fall 1996): 6–8.

13. Cohen, *Gulf War Air Power Survey*, 61.

14. Herbert C. Johnson, *Breakthrough: Tactics, Technology, and the Search for Victory on the Western Front in World War I* (Novato, Calif.: Presidio Press,

1994). On the origins of modern warfare in the effort to overcome the stalemate of trench warfare see Jonathan Bailey, *The First World War and the Birth of the Modern Style of Warfare,* Occasional Paper No. 22 (Camberley: Strategic and Combat Studies Institute, 1996); see also Tim Travers, *The Killing Ground: The British Army, the Western Front, and the Emergence of Modern Warfare, 1914–1918* (London: Allen and Unwin, 1987).

15. On the impact of air power on the battlefield, see Benjamin F. Cooling, ed., *Case Studies in the Development of Close Air Support* (Washington, D.C.: Office of the U.S. Air Force Historian, United States Air Force, 1990). See also Christopher Bowie et al., *The New Calculus: Analyzing Airpower's Changing Role in the Joint Theater Campaign* (Santa Monica, Calif.: RAND, 1993); Eduard Mark, *Aerial Interdiction in Three Wars* (Washington, D.C.: Center for Air Force History, 1994); Thompson and Nalty, *Within Limits*; and Schlight, *A War Too Long.*

16. For examples of the contending schools of thought, see Overy, *Why the Allies Won*; Pape, *Bombing to Win*; Richard Szafranski, "Parallel Warfare: Promise and Problems," *U.S. Naval Institute Proceedings,* August 1995, 57–61; Clodfelter, *The Limits of Air Power*; and Deptula, *Firing For Effect.*

17. Hallion, "Precision Guided Munitions," 11–13.

18. Harlan Ullman and James Wade Jr., *Shock and Awe: Achieving Rapid Dominance* (Washington, D.C.: National Defense University, 1996).

19. John A. Warden, *The Air Campaign: Planning for Combat* (Washington, D.C.: National Defense University, 1988).

20. Stephen T. Possony, *Strategic Air Power: The Pattern of Dynamic Deterrence* (Washington, D.C.: Infantry Journal Press, 1949), 147.

21. U.S. Army, FM-100-5, July 1, 1976, chapter 8.

22. *Discriminate Deterrence: The Report of the Commission on Integrated Long-Term Strategy* (Washington, D.C.: GPO, 1987).

23. Defense Science Board, *Tactics and Technology for 21st Century Military Superiority,* Summer Study Task Force, Vol. 1, Final Report (Washington, D.C.: GPO, 1996).

24. Douglas A. Macgregor, *Breaking The Phalanx: A New Design for Landpower in the 21st Century* (Westport, Conn.: Praeger, 1996).

2

Strategy

Jeffrey R. Cooper

The American conception of how we use air and space power, and, for that matter, how we wage war, must continue to evolve to suit changing circumstances. Few would dispute the assertion that present conditions—framed by a chaotic post-Cold War world of head-spinning technological and political change—are mutating rapidly. Because the United States currently faces no peer-level threats, it is easy to dismiss the opportunity for fundamental change and instead tinker at the margins of what has been a very successful military doctrine and force structure (especially with Operation Desert Storm fresh in mind). However, this situation provides the United States a rare opportunity to engage in essential changes, changes that could be much more difficult to undertake farther down the road when we are already under pressure from a substantial challenger. At a time when serious threats in both the near and medium terms seem unlikely to occur, we are uniquely free to think about the long term, when the United States may face the prospect of a peer competitor. Given the immense technological advantages we now possess coupled with our large, well-trained, and well-equipped forces, we do have the capacity, with timely adjustments to our planned forces, to match any non-peer challenger that might be simultaneously transitioning to a twenty-first century military. We can, therefore, take the modest risk and engage in the sort of transformational change that will truly prepare us for the future, rather than simply making modest adjustments that are suitable for present conditions extended out a further five years. To do this we must create a "new American way of war," one that plays to our greatest strengths in order to ensure a powerful military capability that can function despite future positional and operational constraints.

The United States will remain a strategically defensive and responsive power, reliant on a strategy of power projection. The future American Way of War must recognize the need to seize and maintain the operational initiative while being constrained to a reactive position. The strategic environment in which the nation will be conducting power projection will be more complex and difficult, with peer competitors and robust regional adversaries constituting the principal threats. In view of this environment, U.S. forces will need to shape the conflict even before hostilities begin; these forces must rapidly gather detailed intelligence about the battlespace and then control an opponent's information capabilities, shape his perceptions, and establish operational patterns to seize the initiative as soon as the situation transitions to combat. By dominating the transition from crisis to conflict and then maintaining continuous pressure on an opponent, they will control the initiative and be able to force the opponent into a reactive stance. The new American Way of War must be based on dynamic, adaptive, and responsive operational concepts rather than depending on preplanned responses to predictable contingencies because potential opponents and theaters of operations can no longer be forecast with confidence. The United States must have the ability to conduct forcible entry into a theater and to eliminate any pauses between initial deployment and offensive action. In a world of high-tempo operations, slow, deliberately paced administrative deliberations would leave U.S. forces unacceptably vulnerable to the wide range of threats resulting from the diffusion of advanced weapons technologies. Strategically, the U.S. response to aggression must be less predictable—the traditional sequence (born of the old American Way of War) of establishing a lodgment, building up a preponderance of forces, and only then attacking—is too slow, gives an opponent too much time to mobilize defenses, and leaves U.S. forces concentrated and vulnerable.

The new American Way of War is built on the seamless integration of the different components of U.S. military power, not the independent employment of air and space power. In this new campaign model, U.S. forces must first destroy the cohering elements of an enemy force—his command, control, communications, computers, intelligence, surveillance, and reconnaissance (C^4ISR), strategic attack, and defensive aerospace capabilities—and anything else that allows him to function as a coordinated force rather than merely local discrete units. U.S. forces must then fragment the opponent's large force concentrations into much smaller groups and be capable of defeating the small groups in detail.

In this new construct, the traditional roles of ground and air power are reversed—making ground the supporting element and air the now decisive force. Whereas previously air power prepared the battlefield for decisive ground action, in the new sequence ground forces prepare the battlespace and/or shape the tactical configuration of the enemy, and air power provides the decisive force. U.S. ground forces compel the enemy to concentrate or be defeated in detail, and air power forces the enemy to disperse or have his concentrations destroyed. Thus, the new way of war creates a new role for conventional air and space power, going well beyond existing concepts of air supremacy to air occupation. A new model for air operations can take advantage of the rapid technological progress that has allowed air and space power to overcome former limitations on navigation, communications, target location, weapons accuracy and lethality, and platform vulnerability, to exercise air and space power in new ways. And these new ways hold out the promise of allowing air power to fulfill its true potential in a way only dreamed of by early air power theorists.

A new American Way of War conceived along these lines would be an appropriate and much-needed response to the profound changes that have occurred in the geostrategic and domestic environments. It would keep military preparation flexible enough to meet the as-yet-unidentified challenges of the future rather than tying us to the ephemeral conditions of the present day. It would take advantage of the C^4ISR and strategic attack capabilities that U.S. air and space dominance give us, while remaining consistent, as any model of warfare must, with our fundamental values and national style.

Background

Interest in reexamining the role of air and space power in American military strategy was spurred by the coincidence of three events. First, during the Gulf War the United States demonstrated significantly enhanced air and space capabilities; these were made possible by a series of technological innovations that successfully addressed many of the long-standing constraints that previously limited air power's full potential. Second, the subsequent disintegration of the Soviet Union and the end of the superpower competition resulted in a redefinition of U.S. national security policy and that of our friends and allies as well. Third, the stability in military planning and force posture development that came from a long-term competition with a single, well-characterized opponent also disappeared. The future now appears more uncertain and is likely to feature a range of situations inherently more complex than the confrontations with the Soviet Union and its clients that we had come to recognize and understand. In addition to these three events, and perhaps most important, whatever the enhancements already wrought by the technological innovations demonstrated during Desert Storm, there is now substantial reason to believe that even greater improvements—from the synthesis of knowledge-based concepts to all-weather precision strike, reduced vulnerability and surprise from stealth, and long-range global reach—still lie ahead. It is the central thesis of this essay that exploiting these emergent capabilities both meets the near-term operational requirements and creates the basis for the long-term advantages needed to meet the challenge of a peer competitor as well as regional opponents.

Most current thinking in the United States about the future conduct of warfare and the utility of air and space power occurs within a context of powerfully constraining but usually unstated assumptions about the existing paradigm for an "American way of war,"[1] derived from more than 50 years of experience and habit dating back to before World War II. Perhaps the most important of these factors is the emphasis on overwhelming force and joint (usually meaning four-service) operations, and the phased, sequential nature of our operational plans—factors deeply embedded in our historical images of war from Gen. Dwight D. Eisenhower to Gen. H. Norman Schwartzkopf. Although understandable in the circumstances of the Cold War, this context poses significant problems for U.S. military strategy in the new era.

First, the need to project overwhelming force to great distances creates extraordinary hurdles in the way of prompt, flexible responses to acts of aggression initiated by others; it sets extremely high bars for assessing the sufficiency of force size and strength. Second, a reactive operational posture that unnecessarily concedes the initiative to our opponents only magnifies the difficulties of creating overwhelming military strength at a point of engagement whose time and place is of our opponent's choosing; moreover, conceding the operational initiative is simply inconsistent with expeditionary warfare. Third, our phased, sequential, and deliberate approach to campaign planning sacrifices some of the most potent advantages that the United States possesses for exploiting opportunities created by our forces' growing lethality and effectiveness in an increasingly nonlinear combat environment[2]: individual initiative, flexibility, and responsiveness honed by training and exercises, and unparalleled knowledge of the battlespace fed by our advanced C^4ISR systems. And fourth, this approach inevitably focuses all efforts on affecting the situation at the front, thereby conceding any advantages we might have with respect to operational flexibility and mobility throughout the theater.

Although we could assess the merits of air and space power within the old operational context (and this essay will comment briefly on this approach), it appears worthwhile to examine the proposition that innovative operational concepts based on new technical capabilities for air and space power would allow the development of a new American Way of War that responds far better to evolving constraints, requirements, threats, and opportunities. Needed for this new way of war is, first, a strategic approach and set of operational concepts that restore to the United States both initiative and choice of *how* and *when* conflict proceeds. The ability to control the course of conflict is a strategic imperative—just because we remain a status quo power, on the strategic defensive against upsets to the system, does not mean that we must also concede the operational initiative to prospective adversaries.

Second, by fully exploiting integrated, knowledge-based operational concepts, the United States would be able to seize the operational initiative, control the tempo of operations, and execute offensive operations without demanding overwhelming force. Third, integrating all force capabilities will pose for the enemy a continuous series of tactical and operational challenges that will expose vulnerabilities and open seams for swift and lethal exploitation. And fourth, operations need to be more seamless and temporally coherent, executed without pause in order to exploit our unmatched ability to employ complex continuous operations that exhaust a less capable force. Exploiting the potential of air and space power would do much to support the development of a fresh, effective post-Cold War military strategy grounded, in turn, on more responsive concepts of operation and a more efficient force posture with increased abilities to adapt to the particulars of unpredictable contingencies, as well as being quicker, more forceful, and less vulnerable in the initial response.

The reexamination of the merits of air and space power that is now taking place under the shadow of the Quadrennial Defense Review (QDR) must ultimately address how resources are to be apportioned and applied. For air and space power (far more than for other force elements), this means determining where investments in new technologies—embodied as platforms, weapons, and supporting systems—should be made: to support current capabilities and readiness, to modernize the force over the medium term, or to conduct research and development (R&D) for addressing potential longer-term challenges. But to do this in a meaningful fashion requires that this examination rest on a sound strategic and operational context against which air and space power can be fairly assessed. The question thus arises as to the appropriate role for air and space power within our national military strategy—a fundamental issue involving national objectives and military strategies as well as air and space power competencies.

Altered Environmental Considerations

Unlike the situation during the long confrontation with the Soviet Union, the United States does not today face direct threats to its national survival or immediate challenges to American military primacy—although we should be prepared for idiosyncratic and individually spectacular acts of destruction. We do appear to face, however, an increasingly unstable world and a complex set of geostrategic situations that individually lack not only the vital interests and high stakes, but also the clarity of our previous confrontation with the Soviet Union. Moreover, and not unrelated, the changes since 1989 have come at a time when U.S. forces are shrinking, deployment tempos are increasing, and funding cuts are continuing. Political constraints on the use of force are also tightening due to increased visibility on real-time worldwide news media—tolerance for casualties is low precisely because the stakes in many conflicts are perceived by most Americans as being less than critical.

All of these factors contribute to a decreased ability to maintain political support for drawn-out, indecisive deployments or engagements involving the use of U.S. power for objectives that are neither clear nor necessarily compelling. The American temperament, especially when vital national interests are not at stake (as, by definition, they are not in pursuing enlargement and engagement), is results-oriented, concerned with American lives, and impatient. Allies are likely to be gray rather than white, as opponents (not necessarily enemies) will be gray rather than black; we are likely to find clarity in moral or ethical stances difficult to discern in conflicts that take place over fundamentally political disputes.

It is also imperative to address the issue of the types of wars to be fought, as these provide the critical strategic context and test of relevance for any military force or concept. Bounding our future force planning problem and testing our forces and concepts by two regional contingencies involving second-rate and well-characterized threats hardly serves to outline the real nature of our future military planning environment—one that should be dominated by an open acknowledgment of its inherent uncertainties. This essay assumes that state-dominated conflicts, rather than conflicts with nonstate actors, will in all likelihood continue to be the wars we fight in the foreseeable future. However, this does not necessarily imply that operations between massed mechanized forces over demarcated borders will dominate even this conflict type. We should expect conflicts in less tractable terrain (such as cities, mountains, and jungles); with opponents employing asymmetrical operational concepts and forces, including small, high-tech (possibly mercenary) forces or large numbers of light infantry or irregulars; and involving information warfare, sabotage, murder, terrorism, and other potential threats against the American "rear" as well as our deployed forces. Moreover, we will probably seek more modest objectives, if not political compromises in "limited wars," that may be difficult to equate with the employment of overwhelming lethal force.

Beyond these challenges lie a spectrum of "new threats" that we must be prepared to address with a range of new national security instrumentalities; these include transnational threats from terrorists, organized criminal groups, drug cartels, and purveyors and seekers of weapons of mass destruction (WMD). They could also include threats to international order from natural or man-made disasters—massive refugee flows, complex humanitarian emergencies, state-sponsored genocide or violations of human rights, or major environmental disasters. Some (and hopefully many) of these situations will not occur, but we must be prepared to deal with them even as we maintain our capabilities to deal with state-sponsored conflicts as our primary *military* challenge in the medium term. This spectrum of threats raises questions as to the relevance and utility of the traditional American Way of War which is slow to react, ponderous to move, and presents a large, and potentially vulnerable, presence in the theater. The American people will continue to demand the ability to achieve defined outcomes quickly; we simply need a new and better way to prosecute our objectives—and these create needs for both a set of new military capabilities as counterpart to an expanded national security tool kit.

The current American approach employing overwhelming force may have been appropriate as a means to address major conflicts affecting vital national interests—these occurred only rarely, usually in predictable locations and with long planning horizons. It may be substantially less attractive in an environment in which challenges, even if smaller and less directly threatening to us, are both more frequent and in less certain venues. The reliance on large-scale forces and a phased, deliberate operational approach may be less suited to situations in which opponents may be unable to challenge us directly on the battlefield but have access to weapons of mass destruction, and in which those opponents will increasingly have capabilities to strike at long ranges with ballistic and cruise missiles.

We have entered an environment in which ad hoc coalitions will likely replace standing alliances. Further, immature and unprepared theaters are more apt to be the scene of future operations, emphasizing needs for power projection and expeditionary forces rather than a more static predeployed posture as we had in NATO Europe or still have in Korea. This will complicate relying on postures and operational concepts demanding heavy in-place infrastructures and initial defenses underwritten with substantial predeployed forces supported by extensive cross training, rationalized doctrine, and standardized and interoperable systems, equipment, and procedures. We may also be more dependent on friendly or neutral but not allied nations for en-route basing and staging as well as overflight rights—access that could, under many circumstances, be problematic (as occurred recently with our ally Turkey) in attempting rapid closure to a distant theater. In these immature and unprepared theaters, we will face limited space for bed-down and marshaling of a massive force, and difficulties in protecting the force in the initial stages of a deployment.

In light of these new conditions, other changes further complicate the development of the needed operational concepts for our military forces. Although we may no longer fear an opponent the size or quality of the Soviet Union, sophisticated military systems are increasingly available for commercial purchase, and, as a consequence, threat lead times may be shorter as these items are bought on commercial rather than concessionary terms. With the demise of the Soviet Union, many of its former clients, which are still likely to constitute our future opponents, are displaying less rigidity in equipment, force structure, and doctrine. This universe of potential opponents now displays a wide range of military systems, often bought from our allies, and new doctrinal approaches and operational concepts based on western models and instruction. Further, even with the best of the new intelligence, surveillance, and reconnaissance (ISR) systems, we are likely to have considerably less intelligence about our ad hoc opponents than we had about Soviet and Warsaw Pact forces because few other militaries train and exercise as extensively, are so wedded to formal written doctrine and professional military education, or are as singular a focus for our intelligence collection and analysis efforts as was the Soviet Union and its Warsaw Pact allies. These factors may not change underlying American beliefs or national values and ethos, but they will demand a new approach to how we seek to execute our national military strategy; they should prompt a serious reexamination of our approach to war.[3]

The Old American Way of War

As Russell F. Weigley noted in the *American Way of War*,[4] there are two threads to the American way of war. At one end of a bipolar view of conflict are wars with serious opponents over vital interests (fought usually for absolute objectives). Accordingly, the American people treat these wars with great seriousness, bringing the entire weight and capability of their nation to bear. Even during our Civil War, modern warfare between nation-states was ultimately a test of one nation's strength against that of its opponent—not simply the military forces of one country versus the other's. Throughout this period, military strength was organic to that of the nation, and Gen. William T. Sherman's devastation of the South was intended to destroy its capacity to support continued fighting. In these major conflicts the United States perceived that time was on its side, as it allowed us to marshal and bring to bear our superior national resources and industrial power supported by a national will determined to vanquish evil. The revolutionary challenge from communism backed by the Soviets' rapidly growing military might and international influence convinced many that time was all but irrelevant. The superpower confrontation with the Soviet Union waged with long-range bombers and intercontinental ballistic missiles armed with nuclear weapons brought an unaccustomed sense of imminent vulnerability directly to the U.S. homeland.

At the other pole of the conflict dimension are those hostilities conducted for lesser stakes (i.e., those fought for less-than-vital national interests). In these cases the American people are loathe to shed the blood of their young or spend significant national resources; clearly, the United States has been less comfortable in pursuing limited political goals than more vital interests. But during the postwar period in both types of conflict, air power has played a central role in our conduct of hostilities. In these situations with lower (and less clear) stakes, as we discovered in Korea and Vietnam, lacking unified domestic support and unable to bring our full national strength to bear in these limited conflicts, time was not our ally. A protracted conflict with substantial deployments of ground forces gave our opponents and domestic critics opportunities to undercut the needed public base of support.

Major Conflicts

Almost by definition, future major conflicts in which the United States is engaged will occur "over there" and they will involve regional allies or friends we are protecting from serious external threats. Having been forced to enter a conflict, Americans will wish to defeat the enemy—now evil incarnate—that caused us to abandon the normal state of peace, not simply "bloody the nose" of an erstwhile opponent. Thus, it is worth considering that the conflict with Iraq was a small war masquerading as a big war, but Saddam Hussein reaped the consequences of being demonized as another Adolf Hitler. For conflicts they deem major, the American people bring the entire weight of the nation to bear to prosecute the conflict—now endowed with a strong moral or ethical character and usually in the service of absolute objectives such as "unconditional surrender." Indeed, one might usefully note that World War I, referred to by its European participants as the Great War, was termed by Americans as "the war to end all wars."

In this century, the underlying plan for conduct of major conflicts by the United States has proved to be exceptionally durable; indeed, the model was laid down when Gen. John J. Pershing first took the American Expeditionary Force to France in 1917. This traditional plan is based on the concept of applying overwhelming and unmatchable force provided by the vast American industrial base and logistics support capabilities—burying the enemy under the sheer weight of men and materiél (even without the assistance of technological superiority or inspired generalship). In essence, the plan has four separate, sequential phases:

- Defend, hold, or seize a lodgment—as we did in World War II in both Great Britain and Australia.

- Use the lodgment as a sustaining base in which sufficient forces and logistics can be built up to go on the offensive as well as a base from which the enemy can be attrited (e.g. by air attacks) in order to create an acceptable correlation of forces when the offensive is initiated.

- Having achieved the desired correlation of forces by the combination of buildup and attrition, go on the offensive with large ground forces, supported by air and naval elements, to bring decisive results.

- Having stopped the initial aggression, restored the status quo ante, and stabilized the post-hostilities situation, bring the massive expeditionary force home and demobilize until the next war.

Within this concept, air and space power at the operational level has contributed a *supporting*, even if powerful, set of capabilities to the land forces. Air and space forces were used to prepare the battlespace (providing intelligence and operational preparation as well as air superiority, battlefield interdiction, attrition, and close air support as necessary), and ground forces supplied the lethally decisive force. Air was the forge that prepared the battlefield and the enemy force; the heavy land force was the hammer delivering the decisive blows. When the United States Air Force (USAF) "bought" its independence from the United States Army, the price involved agreeing to a doctrinal and operational legacy tied tightly to a ground fighting construct that continues to this day. Moreover, the resulting allocation of war-fighting "roles and missions" gave special authorities for operations on and over land to army forces and on, under, and over the sea to naval forces. Indeed, a survey of air force doctrine itself reinforces this point as it increasingly paid less attention to strategic offensive operations after Vietnam; none of the missions—air superiority, air interdiction (either battlefield or deep), close air support—describe the offensive use of air on the battlefield as a decisive mission (comparable to the army's decisive assault). The very language of air operations is defined and limited—and to the extent that language limits conceptual thought, the air operations themselves are limited—by the paradigm of land battle. For fifty years, USAF roles and missions have been defined largely in terms of supporting functions even though decisiveness was the core of the argument for air to be an independent force.

In the initial phase of the traditional campaign model, air power played a prominent even if not decisive role. The principal mission here was maintenance of air superiority. Air power provided air cover to protect the lodgment, as was done at Guadalcanal. In the case of the Normandy landings, daylight air superiority was a precondition for attempting to regain a lodgement on the Continent. And in employing lethal counterconcentration and interdiction fires, air power has provided needed respite in direct defense, as occurred in Korea at the Pusan perimeter. Initial air power sent to Saudi Arabia in the days following Iraq's invasion of Kuwait was heavily weighted toward air superiority capabilities. In the second phase, airlift can help to support a more rapid buildup of forces, deploying critical time-sensitive items like command, control, and communications (C^3) and theater missile defense (TMD) units. At the same time, self-deploying tactical air forces (along with long-range strategic bombers) provide the capability to inflict considerable destruction on the enemy, though they require substantial airlift for sustainment in order to significantly attrit enemy forces over time. In the third, offensive, phase, air power affects the opponent's ability to maneuver and respond—a phenomenon seen from the Normandy campaign to Desert Storm—while inflicting significant losses on the enemy from direct attack of its maneuver forces and supply lines. Finally, air power can be a useful element in the final destruction of enemy forces, for supporting the occupation, and for rapidly returning forces home.

The strategic air campaigns waged with conventional weapons against Germany and Japan by the United States and its allies during the second phase of those operations were aimed at destroying industrial war production and enemy morale, primarily as a way of attriting the enemy's military potential. These campaigns were fundamentally directed toward weakening the operational military capabilities of the enemy—rectifying the balance of forces—rather than defeating the enemy by direct homeland attacks, as had been advocated by some air power enthusiasts.[5] The naval air forces, similarly, were originally developed to substitute for naval gunfire as the principal striking power of the fleet directed at naval targets. This strongly held doctrinal orientation toward operational targeting led to a major confrontation in the late 1940s and early 1950s with the new U.S. Air Force over the purpose for and/or application of strategic air power.

Less-Than-Major Conflicts

A second thread to the American way of war applies in those cases in which neither the stakes nor the opponent appear substantial—and, therefore, in which we are loath to put large numbers of American lives at risk in service of limited political objectives with no clear outcome. We prefer our conflicts to be short and decisive, or at least of limited duration; we do not like open-ended military commitments, and we have tried to limit our long-term constabulary operations except in support of formal allies. As a consequence, we have usually sought to avoid limited wars requiring extended campaigns in pursuit of these limited objectives, and air power has seemed to be a natural even if not fully effective answer.

This thread, employing punitive raids (or "strikes" in air parlance), conceptually goes back even further—to the French and Indian Wars and Thomas Jefferson's ultimate treatment of the Barbary pirates. In its modern postwar guise, this was the raison d'être behind bombs on Libya, Lebanon, and Serb military facilities, as well as, more recently, cruise missiles on Baghdad and assorted Iraqi air defense sites. Although substitution of modern technology can minimize the numbers of Americans put at risk, the downside is the lack of overt, evident commitment. Not applying massive force but still employing the concepts derived from both "big war" and nuclear doctrine to small-scale conventional actions may be a serious mismatch of means and ends. And so far, the track record for this use of air and space power is effective but limited as a deterrent against future threats of aggression. Moreover, the United States has often mischaracterized minor conflicts in attempting to garner public support and has wound up prosecuting these as major wars with all the attendant baggage. Although this may yield some benefits in the short term for support and prosecution of the conflict, it may create longer-term problems concerning expectations and perceptions from the mismatch of objectives and forces.

Decisive Air Campaigns and Deterrence

As was demonstrated during World War I and reinforced during World War II (especially on the Eastern front), if unhindered, the immense productive capacity of modern industrial nations could produce and equip more military forces than even the lethally effective modern weapons could kill on the battlefield. In response, early air power theorists such as Giulio Douhet saw that air forces could provide a way to attack the industrial underpinnings of a modern opponent's military without having to fight through the enemy's forces to reach the "strategic rear." However, the technical limitations of air power armed with conventional weapons constrained the ability to fulfill this promise; but even so, in World War II air power, by weakening the industrial base,[6] did allow a decisive victory to be obtained against Germany, unlike in World War I.

The atomic bomb enabled air power supporters to argue successfully for a new approach to war based on direct attacks against the enemy's critical homeland target structure as a strategically decisive capability, as opposed to attacking his fielded military forces. A separate air force (with Strategic Air Command, or SAC, as the jewel in the crown), independently employing atomic-armed strategic air power against the enemy homeland, would not merely weaken his military forces but bring decisive force and shock to bear in order to achieve strategic victory directly from these attacks. Indeed, Eisenhower's New Look military, exemplified by the Pentomic divisions and atomic-armed tactical air forces, continued the thrust to substitute atomic lethality for force structure across the board.

After the collapse of the Soviet Union, however, the strategic nuclear mission no longer held primacy in American military policy. Was there a case, then, for an independent air campaign based on a capability to achieve independently decisive results by attacks with conventional weapons on the opponent's infrastructure and war-making potential? If strategic nuclear attack can no longer underwrite the case for a war-winning air campaign, this case must, therefore, rest on other ground—one of two related but distinct arguments for the *decisive employment* of air power. Past failures to construct a solid foundation for a war-winning approach, absent nuclear weapons, do not give strong encouragement for success in the future unless significant changes in strategic approach and operational concepts are adopted. The first argument is that air power could be separately employed, as in the Gulf War, in a "strategic air campaign" focused on the opponent's "strategic centers of gravity"; indeed, this is a conventional cognate to the nuclear planning of the 1950s through the 1970s.[7] Its utility and effectiveness as a war-winning capability by itself rests on an ability to limit an adversary's options and, in many cases, to shape enemy behavior through critical damage to leadership, control mechanisms, infrastructure, and industry—many of which are increasingly intertwined with civilian society. Notwithstanding the enhanced precision and lethality of sophisticated brilliant weapons that levied lethal destruction on Iraqi war-waging and supporting capabilities, it has not been convincingly demonstrated, even during the Gulf War, that this type of widespread destruction represents—by itself—a war-winning capability for achieving decisive strategic victory. Moreover, even if it were to be technically feasible and sufficiently effective, a doctrine that required that degree of lethality and destructiveness could be difficult to exercise for political reasons, both domestically and internationally, in less than in extremis circumstances.

Alternatively, could modern air power be directed at engaging and decisively defeating operational targets—the enemy's fielded military forces—as the focus for an air campaign.[8] This option assumes that operational success against the opponent's military forces can produce strategic victory. Indeed, the attractiveness of this concept now goes well beyond the U.S. Air Force alone. Other services also see in this vision of the Revolution in Military Affairs (RMA) an opportunity to be the decisive force by bringing their long-range lethal firepower to bear on enemy forces engaged throughout the battlespace. Some in the U.S. Air Force, as well as in other services, continue to support the concept of overwhelming force but argue that this paradigm requires substitution of the newly lethal, concentrated long-range fires (by air and space power in the case of the U.S. Air Force)—the new version of overwhelming force—for the traditional large-scale ground forces as the central element of the American plan for victory in the theater. In fact, some argue that—especially with the substantially enhanced effectiveness of modern air- and space-based ISR systems to provide more precise and timely indications and warning—mobile, responsive, and lethal air and space power, executed through an independent operational "strike" campaign, could prosecute this overwhelming force model by itself.

There does not appear to be substantial evidence that these promises for an operationally decisive capability, even supported by sophisticated platforms and weapons, can be fulfilled with strikes by air power alone. If neither of the two arguments for a separate air campaign as a war-winning capability proves persuasive, then the case becomes far stronger for exploring the merits of air and space power as a decisive element in the context of a truly integrated joint campaign rather than as an independent instrument. This does not mean, however, that air and space power must inevitably be adapted to a supporting role.

These two alternative approaches to the application of air and space power also carry significant implications for how we conceive of and implement deterrence. We have always recognized that effective deterrence required that an opponent understand that we possessed both *credibility*—his belief in our will to unleash the deterrent force—and *capability*—his belief that we could effectively execute our deterrent instrument. What was less apparent as a requirement because it was implicitly accepted by almost everyone was a third crucial element, a common appreciation of *consequences*. In the era of nuclear deterrence, we all understood that fundamental national survival was at issue if large-scale nuclear strikes were executed. The potentially lowered threshold for employing conventional forces, especially if they are discriminate and promise reduced collateral damage, should help to underwrite U.S. credibility in threats to use force in retaliatory response to aggression. The evident capacity of the force to penetrate even sophisticated air defenses and its lethality when employing advanced precision weapons makes a prima facie case for air power's capability to carry out retaliatory punishment. Despite these two evident advantages, however, it is not certain that even the most lethal conventional attacks carry a perceived threat to an opponent's national survival; therefore, without such a "fatal vision" to underscore the retaliatory consequences of aggression, conventional "strategic attack" (i.e., in the absence of nuclear weapons) may not constitute an effective a priori deterrent. But in conjunction with other consequences that could affect regime survival, the potential for fatal consequences of initiating aggression could be demonstrated.

The first course, focusing on specific attack of enemy homeland targets, thus does not appear to provide a solid case for an independent air campaign based either on deterrence or war-winning effectiveness. With respect to the second course—threatening the direct destruction of an enemy's military forces and thereby preventing the opponent from meeting his objectives—it does not appear that a convincing case has yet been made that the independent use of air power can destroy his military forces; or, even if air power could destroy the enemy's projected military forces, that this capability would necessarily translate into effective deterrence of the act itself. But this concept does properly reopen the larger issue of alternative paths to reestablishing effective deterrence if strategic targeting is not effective, whether by destruction at the operational level or by denial of the adversary's military objectives. Even if air power alone may not be sufficient, it is worth exploring whether integrated forces could create an effective deterrent.

New Opportunities for Air and Space Power

Air power has been a key factor in American military power since its incorporation into the defense establishment during World War I. Air power contributed substantially to victory in World War II in both the European and Pacific theaters, was a crucial element in staving off early defeat in Korea, supported U.S. and allied forces in Vietnam, and underwrote American strategic policy in our decades-long confrontation with the Soviet Union. But, whatever the theoretical merits put forward by air power advocates for its overwhelming military effectiveness, in the past air operations were dogged by severe limitations—usually technical, but sometimes of strategic thinking—that historically prevented air power from being a decisive instrument. Recent technological advances, both enabling and coupled with a revised strategic approach, innovative doctrine and operational concepts, and restructured combat organization may have finally tipped the balance in favor of air power. Perhaps most important, the old way of employing air (and increasingly space) power may have outlived its ability to exploit most effectively the underlying capabilities made possible by sophisticated new technologies and, therefore, to fulfill air power's true potential.

Technical Challenges Addressed by the Working Group

Navigation. Until the advent of the modern inertial navigation systems (INS) and the space-based global positioning system (GPS), it was difficult for air forces to know where they were and where they were going with accuracy and timeliness, especially at night and in bad weather. Modern geolocation systems create a navigation framework that enables forces to locate both themselves and the enemy within a common coordinate system and to do this for both static and mobile targets under almost all environmental conditions.

Intelligence, target location, and identification. Prior to modern overhead and aerial surveillance and reconnaissance systems, finding and identifying targets, including the critical aim points, was an extraordinarily difficult task even for static targets. And without the new real-time battlespace surveillance systems (such as the Joint Surveillance Target Attack Radar System—JSTARS), together with supporting C^4 systems, finding moving targets in real time on the battlefield so they could be promptly attacked was nearly impossible. Indeed, the basic concept behind the existing air campaign model and the air tasking order (ATO) methodology was to focus the air effort by attacking a list of fixed static targets; this could be done by preplanning missions based on previously located and identified targets.

Reliable communications. New communications technologies, many of them space-based, now promise air forces and their commanders the ability to communicate reliably while in flight, in both directions, securely, and with substantial bandwidth—regardless of geographic location. These new technologies allow shared situational awareness not only among sortie elements but also between echelons at all levels. They allow passing of high volumes of real-time data or critical en-route updates of targets or defenses. These technical capabilities enable planners to alter many of the traditional operational concepts on which modern air power was based. Moreover, they remove many of the limitations that prevented effective real- or near-real-time command and coordination of complex air operations, thus allowing engaged air assets to achieve desired effects with reduced vulnerability.

Weapon accuracy and lethality. Modern weapons, whether equipped with autonomous seekers, accurate geolocation capability, or off-board terminal guidance assistance, together with advanced warheads that significantly increase lethality compared with bulk explosives, make accuracy and lethality far less range dependent than in the past. These new capabilities give long-range indirect fires the kind of effectiveness traditionally found only in short-range direct fire systems, a feature that has significant implications for the composition of strike packages, numbers of supporting forces, logistics, and the overall conduct of air operations.

Survivability. Again, recent technical developments address critical problems that have plagued air power since the development of radar-aimed gunfire and antiaircraft guided missiles and the introduction of effective integrated air defense systems. The combination of stealth, advanced electronic countermeasures (ECMs), and effective suppression of enemy air defense (SEAD) begins to return an important measure of survivability for an entire air campaign, not just quick sorties into "denied areas." Moreover, stealth has also been found to be an important contributor to weapons accuracy and lethality. Although air forces could always threaten to place some weapons on targets despite severe attrition, an ability to accomplish the entire spectrum of critical air missions with acceptable losses throughout a campaign underwrites the ability to wield air power as a decisive instrument. Further, the ability to minimize the risk of aircrew losses (either through death or capture) increases the usability of air power as a tool for national security decision-makers in low-level conflicts.

Surprise. In addition to its tremendous benefits for aircraft and crew survivability, stealth returns to air and space forces the critical ingredient of surprise that had been eroded by radar and long-range early warning systems. Like the intercontinental ballistic missile (ICBM) in the early 1960s, stealth fundamentally alters the utility of air power until effective detection and countermeasures can be developed. The element of surprise not only increases the shock of an attack but can also help to restore the operational initiative to American decision-makers and military planners responding to acts of aggression.

Finally, it is important to recognize that, although each of these innovations is individually impressive and important, when integrated they form a synergistic ensemble of amplifying capabilities that change the character of air and space power as it has been known. For example, the combination of information, precision, longer range, and stealth could go far to obviate the need for traditional force packaging, thereby greatly reducing the forces required to carry out specific operational tasks as compared even with Operation Desert Storm.

Operational Considerations

The Allied air campaign, conducted as a precursor to the Normandy campaign, not only achieved air superiority, but also succeeded in barricading the region from German troop reinforcements or logistics support. And once the invasion began, the same air superiority allowed the Allies to control the ability of the Germans to maneuver and respond to Allied operations. But, as illustrated during the Battle of the Bulge, even under conditions of operational air supremacy, air could dominate the ground only during the day and in clear weather with older technology that lacked modern sensors and computer processing.

The Gulf War demonstrated that, even though not all problems had been solved, substantial progress had been made in overcoming the critical technical constraints on decisive air and space operations. As the opening night air operations showed, combined arms air power enhanced by stealthy operations could penetrate and destroy even sophisticated integrated air defense systems (IADS). Air and space capabilities could dominate the battlefield not only during the day; in some cases, sophisticated target location sensors actually performed better at night. However, bad weather still adversely affected critical air operations.[9] Stealth, despite its value against radar air defenses at night, still leaves open the question of how to conduct effective operations, including long-loiter and persistent operations, during the day or against proliferated man-portable infrared missiles. Further, despite even multihour endurance missions over enemy territory, U.S. air forces were not very successful in hunting Scud missiles. But, though fleeting mobile targets remain among the most difficult problems to be solved, technical advances in sensors, information processing, and quick-reaction weapons suggest significant prospects for improvement. Space-based capabilities, many originally developed for strategic missions, added immeasurably to the effectiveness in prosecuting tactical battlefield operations; nevertheless, there are concerns about survivability of space-based assets and the overall dependence of U.S. forces on sophisticated but potentially vulnerable C^{4}ISR systems.

Resolving these troublesome operational issues and constructing a new model for air and space operations would create a powerful tool for military commanders, with immense strategic and policy implications for national command authorities—as well as for our potential opponents. The technological advances discussed here will enable the development of a new set of operational concepts that can support a fundamentally different paradigm of how America fights its wars. This paradigm will change not only the conduct of future operations on the battlefield, but, perhaps more important, our overall strategic approach to the employment of military power as well. The demands for the exercise of military power placed on a single, dominant, global superpower are entirely different from those required by a nation locked in a titanic, long-term mortal competition; as a result, the United States must reshape its approach to military strategy. Therefore, these new needs provide a different context for how we apply technological innovation to air and space power.

The Role of Technological Innovation

Relying on qualitative advantages from sophisticated technologies to offset the quantitative advantages possessed by the Soviets dominated our thinking about technology throughout the Cold War and given our perceptions of the Soviet threat, we pursued three distinct approaches. First, because we believed that we needed to maintain a qualitative edge, we thought that we had to make incremental improvements in military systems across the board if we were to prevent gaps that the Soviets could, and would, exploit. Thus, for example, we planned new generations of aircraft that could fly faster, higher, and farther; roll quicker; and turn tighter. Second, we looked for innovative applications of technology, not simply direct counters, that could produce revolutionary impacts and enable new operational concepts. The United States developed the LANTIRN and LONG-BOW targeting systems, for instance, to enable fixed- and rotary-wing aircraft to locate, identify, and destroy fleeting ground targets autonomously, with minimal pilot intervention even at night or in bad weather. And, we sought fundamental scientific or technical breakthroughs, like stealth, that would produce revolutionary effects and leapfrog Soviet systems and capabilities, and then doctrinal and operational concepts as well.

Today there are no state-based threats to our military superiority in either quantitative or qualitative terms. Nevertheless, individual systems already fielded, like the SA-10 surface-to-air missile (SAM) stress U.S. systems; these could cause significant losses when deployed in quantity by states wealthy from oil or other revenues whenever they choose to do so—and by others when they can afford them. With no mid-term challengers that seem capable of mounting a serious threat to our technological advantages, we have been denied choices and flexibility over the past half century in how we apply new technology and to what objectives. We no longer need to use our technological advantage to offset quantitative deficiencies. The question of how to apply the impressive new technical capabilities of air and space power to military needs remains open in the absence of a threat-driven imperative that did not appear to offer any obvious weaknesses.

Our future opponents are not simply smaller versions of the Soviet Union, however. All show significant gaps in capabilities that innovative operational concepts, enabled by the technological advances, could exploit to avoid confronting enemy strengths head-on. Although we should remain alert to the dangers posed by asymmetric strategies and capabilities, we also must not ignore the opportunities they present to us to create a warfare paradigm suited to our own needs. Thus, there do appear to be strong reasons for seeking to use these new air and space capabilities to transform our conceptual military approach rather than merely to make marginal improvements in system performance or to enhance how we conduct our current tasks. These are risks we can now afford to take because there is no opponent who could effectively exploit any opening or vulnerability presented by our deliberate transformational path.

The list of technical advances and innovations for air and space power is impressive, and there is a long list of potential paths and investment options for exploiting these advances, as the application of technological innovation could support very different objectives. But fundamentally these developments can be grouped into three categories: enhancements to current systems and forces in order to improve current missions; modernization of major air power systems, either to maintain technical advantages or to expand the types of missions conducted; and R&D for longer-term concepts that could offer more than incremental capabilities and allow us to pursue new strategies with different operational approaches. Simply appliquéing sophisticated new technical capabilities on old equipment, operational concepts, or organizational structures in service of outdated strategic concepts appears to ignore the lessons of effective technology application. The greatest gains occur when new technology enables and is embedded in a synergistic matrix of improved systems, revised doctrine, innovative operational concepts, and restructured organizations all designed to accomplish new objectives. Indeed, the application of new technology often follows a three-step course over which time the objectives it serves change dramatically.[10]

Initially, technology is often used to simply improve performance and meet already defined needs in executing a defined existing, and often specialized, task—that is, to do a current hard job better, cheaper, or more efficiently. Thus, brilliant air-delivered weapons provided greater lethality against difficult-to-hit targets (such as bridges) for each sortie and for overall effectiveness of the air operation. The second stage of technology exploitation often results in a dramatic increase in the amount the new innovation is used because the performance advantages it enables make it relatively more efficient or cheaper compared with other products or services—that is, a "superior good." And as its application base spreads to users who develop new or different ways of performing a wide variety of tasks utilizing it, this will have significant displacement effects of existing products or processes resulting from its intensive exploitation as its ability to substitute for other methods or products is widely recognized. The widespread adoption of precision-guided munitions (PGMs) (compared with the Gulf War in which PGMs made up only about 8 percent of all air-delivered munitions dropped) for future planning demonstrates the extremely broad spectrum of targets that PGMs appear best suited to service. As weapons employment moves toward an all-PGM force, the entire size and composition of air operations will be altered.

Finally, the third stage often involves a transformation that transcends the narrow set of new technical capabilities or task options as users reformulate their objectives, or at least their conduct, as a result. Application of superior technology through the first two stages did not alter the basic nature of how air and space power has been employed: as a supporting instrument in episodic doses. In this transformational stage, the technological advances outlined above, when used together, could allow air and space to create a new model for future air operations—such as seizing control of the medium to prosecute decisive operations with continuous air and space capabilities—thereby altering the character of all combat.

The global positioning system can also serve as a useful example of this progression through these stages. Initially, GPS was conceived as an adjunct and enhancement for military aircraft inertial navigation systems, especially for long-range strategic missions in which INS errors could accumulate over many hours.[11] But due to its low cost, small size, and high precision, GPS was adopted widely by civil users as well as incorporated in a wide range of other military equipment such as land navigation systems; it has progressed far beyond the uses originally conceived for aircraft navigation systems. Uses such as individual bomb guidance or package tracking could not have been implemented with even miniaturized mechanical INS systems due to cost and size constraints. In the case of GPS, the third-stage transformation has involved the creation of an implicit single worldwide geographic reference system (an earth-centered grid) to which any object's position can be related in real time. It has not only transformed navigation by air, sea, and land; it has also changed classic accuracy relationships for weapons that were distance-related. Moreover, in building a world in which the position of everything is precisely known, it has created fundamental new business and scientific opportunities for tracking packages, finding stolen cars, and measuring continental drift.

A New American Way of War

Times are changing ever more rapidly and the nature of the national security endeavor is evolving accordingly. Whatever the specifics of these unpredictable changes, the United States intends to retain its leadership position and its overriding military advantage. We could attempt to meet these demands by making marginal improvements and "tinkering at the edges" to systems and concepts, but a bolder approach is to confront directly the nature of our altered circumstances and create a new American way of war. In constructing a viable military strategy throughout the Cold War, we worried about three key factors: the nuclear overhang and controlling escalation, the enemy striking first and then the North Atlantic Treaty Organization (NATO) fighting outnumbered; and time being on our opponent's side. Today, each of these circumstances has been significantly transformed, and though we may have taken the first change too much to heart, it is not clear that we have fully internalized in our thinking the implications of the second and third. Whereas we may not have to worry about escalation of minor conflicts to global confrontation, we may worry too little about escalation to use of WMD by our opponents. The change in the second factor should allow us to explore the use of asymmetric strategies and operational concepts to exploit our opponent's weaknesses rather than continuing to rely on massive force amplified by technological effectiveness. The third factor suggests that current emphases on responding immediately and promptly halting acts of local aggression may need to be reexamined, especially in light of historical experience.

Finally, there is little evidence that we have understood the implications of opponents who fight without the protection of superpower sponsors; under the old rules, we never sought "rollback" but merely return to the status quo ante—thereby limiting our opponent's potential losses. In the new circumstances, our opponents should not be assured that their table stakes are limited to the loss of the objective they sought, but crossing the line implies unlimited liability—at least of regime survival. Thus, although we may be uncertain as to the rationality or exact frame of reference of future opponents—and therefore the details of their strategic calculus—the role of fear that unspecified but potentially unacceptable consequences will occur should not be underestimated.

Potential concepts for a new American way of war must address not only the inconsistencies that have already arisen between the traditional approach and the changes in the geostrategic and domestic environments noted above, but also potential new challenges (especially threats to our most vital strategic interests) that we have yet to identify but would be foolish to ignore. It would be the height of arrogance to assume that over a 20-year period no serious challenger to our interests will appear. Yet we must remain consistent with our fundamental values and national style whatever the means through which we choose to pursue our interests. Recognizing that U.S. forces will continue to shrink, we must make effective use of our reduced resources in order to maximize both deterrent and combat power; therefore, an approach that integrates diverse military capabilities into a single coherent joint force that amplifies their inherent power can contribute enormously to maintaining America's military advantage.

Both previous dominant superpowers, Rome and Britain, understood that they could not afford to be everywhere or respond immediately, but they also understood that certitude of their eventual and dominant response was essential to maintaining peace on the frontiers. We now have the flexibility to choose the timing and pace of our responses consistent with constructing an effective but sustainable military capacity through this period of no overt peer challenge; the invasion of Kuwait was neither deterred nor halted, but it was reversed. And we are likely to face more threats that cannot be deterred but must be confronted and corrected on the battlefield.

This is not to argue that such a capability would be a panacea for all of America's challenges abroad. Some problems do not have military solutions, and fewer still are amenable to the decisive exercise of air and space power alone. However, air power capabilities that go beyond traditional air and space supremacy would enable rapidly responsive, highly adaptive operations on a worldwide basis. These would provide American decision-makers with a substantially increased range of effective options to address our future strategic challenges.

The New Requirements

As the only remaining superpower, the United States retains a set of global interests and worldwide commitments shared by no other nation. Consequently, we must have the capacity to be engaged in any region, yet we cannot ignore the complex constraints on our freedom of action, both domestically and internationally. We will need the capacity to monitor events around the globe, shape worldwide developments by our influence (however exercised), prevent and deter open hostilities, and, when necessary, successfully prosecute military missions ranging from peacekeeping to major contingencies without a World War II-style national mobilization. The American people are likely to demand that U.S. forces be able to respond to these worldwide contingencies promptly, achieve decisive results, avoid becoming bogged down in a lengthy campaign of attrition, and return home for Thanksgiving turkey (or at least Christmas goose). Complicating these needs, unfortunately, is that in only a few circumstances will we find sufficient clarity of objectives or a Manichaean enemy to allow us to pursue absolute objectives such as unconditional surrender with unconstrained force. Given the rapid, even revolutionary, advances in

C^4ISR technologies and our domination of those technologies, a new American way of war should increasingly seek to move information instead of physical materiél and replace the need for overwhelming force with decisive knowledge-based action, enabled by global ISR capabilities, that can be calibrated to the actual challenge.

The new American way of war must shift toward dynamic, adaptive, and responsive operational concepts rather than relying on preplanned contingency responses designed for more predictable situations that occurred in a more stable environment with fewer chaotic changes. These new concepts must support responsiveness without the concomitant need for predeployed forces or pre-positioned materiél exactly because in many circumstances we will be unable to forecast these needs before the situation erupts. (Moreover, even where we foresee the need for these preparatory measures, in the absence of formal alliance commitments our potential hosts may simply not agree to the measures until the wolf is at the door.) But once the threat begins to materialize, we need the capacity to begin responses that underwrite and underline both our ability and our intention of ultimately achieving a decisive outcome—one that does not simply return the situation to the status quo ante. Potential aggressors need to understand that these are not risk-free exercises and that a real downside exists beyond not achieving their objective.

Once the "line" is crossed by the aggressor, the options and pacing should be ours despite the likelihood that our opponent will initiate the conflict at the time of his choosing. Instead of responding to an adversary's choice of course and timing of action, and therefore conceding to him the operational initiative, a new American way of war must retain and control the initiative in keeping with the nature of expeditionary warfare—and this is likely to demand capabilities for something other than simply force interposition and head-to-head combat. The United States must also have options for preemptive actions such as spoiling missions or even large-scale decisive strikes in order to give our decision-makers the ability to retain control of the course of the conflict. Operations should be conducted within the entire breadth and depth of the theater—not in series, but in closely coupled parallel operations; moreover, operations aimed at strategic, operational, and tactical objectives should be conducted simultaneously in order to seize the initiative, maintain the pressure on the opponent, and give him no pause or sanctuary to catch his breath. Each campaign must be designed from the start as an integrated, joint force, combined arms effort, in order to exploit our inherent advantages.

To be capable of successfully confronting unpredictable acts of aggression, the new American way of war must also provide a capability for "forcible entry" instead of administrative insertion into a theater. Our new campaign model must also incorporate the capability to operate in "immature," unprepared theaters without extensive preparation or host nation support; further, we must be prepared to conduct those deployments in nonpermissive environments. We should not count on having unhindered airheads and ports to receive our forces prior to the start of a conflict, nor should we depend on undegraded air bases in the local area of operations as the underpinning for an unchallenged air campaign. The new American way of war must plan for a more seamless approach, without pauses once deployment to theater begins, in order to both minimize exposure of the forces to attacks on static positions, as well as to deny the opponent an opportunity to interfere with the critical domestic political support for these operations. It would be dangerous to assume that the advantages that we had during the Gulf War will be given to us again. We need a campaign model that does not take these as givens and can secure necessary operational facilities under fire if necessary.

A New Campaign Model

New operational concepts based on our advanced technical capabilities should enable us to create a more effective military strategy—indeed, a new American way of war. Although some opponents in some operating environments may be able to conduct their operations without air and space superiority, or even without any effective air and space capabilities, for U.S. forces, air and space superiority is the sine qua non for all else. Therefore, dominance of the aerospace domain will be essential in future operational planning, and this must become a key requirement in all future investment decisions. Such dominance, however, is not just a product of air superiority aircraft; it is a combined arms problem that involves addressing all threats and sources of interference with our needed air operations. As a historical reminder, it was Gen. Ariel Sharon's armored brigade thrown across the Suez Canal that destroyed the Egyptian air defense system during the 1973 War and allowed subsequently unfettered Israeli air operations. Dominance in air and space power can totally alter the conduct of operations beneath. Therefore, this capability—which we could not depend on against the Soviets—now offers us the opportunity to reshape how we employ our ground and naval forces against all opponents.

A new campaign paradigm built on these concepts offers one path to minimizing vulnerabilities, increasing concurrency, shortening the overall duration, and substantially enhancing the combat effectiveness of U.S. military forces by matching our strengths against the opponent's weaknesses. This campaign model is designed to avoid the necessity of directly confronting an opponent's "peaks of advantage": in most cases, these are his massed forces or his weapons of mass destruction. It is especially intended to prevent the enemy from gaining an advantage from offensive or defensive postures for deliberate, prepared operations. The degree of concurrency among the operations is high, the transitions are seamless, and the overall duration of the campaign is extremely compressed. This model tries to avoid the "set-piece" nature of the existing predictable model in which we gain a lodgment and then attack out of it, thereby offering the enemy time to establish his defenses and employ tactical missiles (potentially with WMD) before we are capable of effectively defending ourselves.

It is crucial that the campaign not wait for the opening of overt hostilities to begin operations that shape the battlespace and seize the initiative; operations that are essential to gaining dominance of the conflict before it actually occurs. In fact, this need has always existed, but, in the past, the focus of preconflict preparations and mobilizations has been on equipment and personnel readiness, positional advantages, and logistics buildups to support the required correlation of forces for either attrition or overwhelming force operations. In the new model, pre-conflict operations are designed, instead, to "control" the opponent's information capabilities, to shape his perceptions, and to establish U.S. operational patterns so that American forces are prepared to transition seamlessly and seize the initiative as the situation moves from crisis to conflict. Information warfare (IW), in addition to the technological advances noted above, provides a powerful set of tools to assist war-fighters in these pre-conflict tasks. Especially in light of concern over and potential constraints on preemptive actions by our allies and friends, a tool kit of nonlethal but effective measures is a welcome addition to military commanders' options. Air and spaces forces, particularly because of their rapid responsiveness and global reach, will play critical roles across the entire C⁴ISR and IW domains.

A key emphasis of the new campaign model is to dominate the transition from crisis to conflict without necessarily having to strike the first blow, not simply preparing to assist in entry of the forces needed for decisive combat in subsequent phases of the operation. Information warfare and perception management techniques can play an important role with allies, friends, enemies, and domestic constituencies in establishing objectives, creating thresholds, and controlling the critical transitions. In the near term, lacking a peer competitor and with less concern for escalation, the United States should be prepared to adopt a more aggressive posture and, during the opening moves, strike with maximum force on transition to open hostilities in order to seize "battlespace dominance." Once these forces are introduced, they will maintain continuous pressure on the opponent in order to force him to *react* to our plan and tempo of operations. Thus, the initial operational moves must set the terms for the entire campaign.[12] In this context, two requirements appear to be of the highest priority for ensuring "battlespace dominance": establishing a dominant position in surveillance and reconnaissance as well as air supremacy.

Unfortunately, under the new geopolitical conditions, we cannot afford to provide detailed coverage and local knowledge of all potential operational areas; the focus must be on the most likely or important areas of potential operations. As crises begin, however, we will need to initiate measures to intensify coverage that is reversible without excessive cost.[13] Thus, deployable airborne systems such as high-altitude, low-observable and high-altitude, long-endurance (HALO/HALE) unmanned aerial vehicles (UAVs) and manned surveillance assets appear to be particularly useful in this regard. But because there is no guarantee that secure, survivable, in-theater ground-based facilities will be available at the onset,[14] it would appear warranted to launch critical surveillance and reconnaissance (S&R) assets such as HALE UAVs from carriers or other naval assets as well as land bases and, similarly, to operate other surveillance assets from maritime or other secure sanctuaries. In turn, this implies that the United States Navy will have to make provision on-board for the critical supporting facilities (such as high-bandwidth antennas, computer processing, and communications equipment for dissemination) needed to exploit these systems not only for its own assets but to support integrated joint capabilities and concepts such as the joint forces air component commander (JFACC) afloat. The capability to establish effective, responsive, and survivable S&R will be an essential element of a successful operation, and it is essential to develop deployable postures that will not be tied to vulnerable fixed facilities in the theater within range of the opponent's missile and unconventional delivery means.

Once conflict actually begins, three critical, largely concurrent operations are conducted as part of the new campaign. The first critical operation is an integrated effort to destroy the opponent's essential C^{4}ISR capabilities that allow him to command and control his forces. This will enable us to negate the adversary's capabilities to extend or maintain his strategic offensive reach, and to significantly degrade his critical defensive aerospace capabilities. This operation is designed to force the enemy into a "local" mode by disrupting the critical command linkages that allow the leadership to exercise "global" coordination; executing them effectively may become more challenging as opponents shift to less nodal organizational structures and facilities. Even in the absence of an overwhelming advantage overall, the United States—enabled by its superior knowledge-based concepts of operation—could respond inside the enemy's decision and action cycle and produce overpowering local superiority time and time again. The new campaign model recognizes that winning at least air dominance and fracturing the cohesion of the opponent's forces are essential missions that must be accomplished soon after hostilities begin and before major surge and sustainment forces are placed at risk in the immediate theater of operations. Rather than building up forces to produce a suitable "correlation of forces" for an attrition-style campaign, this model focuses on destroying the critical cohering elements and then fracturing the opponent's forces so that the overall force correlation is meaningless. Perhaps most important, these actions, when combined, are intended to give U.S. forces the initiative in the timing and tempo of subsequent operations by driving the opponent into a reactive mode, continually pressing him, and not allowing him to regain command of the combat process, as well as to help us achieve battlespace dominance throughout all mediums of combat operations. We will need the capacity to do this against future opponents that are more agile and effective than Iraq during Operation Desert Storm.

Having destroyed the cohesion and structure of the opponent's plan, the second critical operation of the campaign, begun without pause or noticeable transition, is designed to "fracture" and "fragment" the enemy's main forces, like Gen. Heinz Guderian's 1940 Ardennes offensive—not, as in the old model, merely to attrit them in order to set the stage for a decisive force-on-force combat between massed armies. In the third and final critical operation, American or coalition forces will exploit the opportunities created by fragmentation and defeat the opponent in detail, without having to face concentrated forces or fires or attempting to force a single decisive action. Under these conditions, old images of linear combat and massed formations need to be replaced with an understanding of the dynamics of widely dispersed combat operations conducted in a sparsely populated battlespace by fleeting targets.

This new campaign approach also explicitly recognizes a fourth operation—recovery and reconstitution of forces. In a period in which conflict may not be a singularity—that is, the United States is not preparing for *the* war between NATO and Warsaw Pact forces—it is likely that we may be faced not only with potentially simultaneous contingencies, but also with situations in which conflicts are repetitive. Therefore, campaigns should be designed to minimize difficulties in extraction while maintaining a high degree of effectiveness. For example, substituting short-term functional disruption for physical destruction where feasible and not operationally costly has considerable postconflict advantages. In most cases where the United States has been victorious, we have ultimately paid much of the cost of restoring our former opponent as a functioning, viable society. This is not likely to change.

New Operational Concepts—Beyond Air Supremacy

The new campaign model is based on a fully integrated operation with advanced C^4ISR capabilities as its backbone. It calls for a substantially closer degree of coordination and synchronization among all operating elements: sharing a common tactical objective, employing common doctrine, operating in synchronization at the tactical echelons, and providing mutual support. Diverse sets of force capabilities (such as air and ground) are directed at a single tactical objective and are coordinated in space and time to achieve a decisive tactical result that can be exploited in this nonlinear environment to produce significant operational victories. This integration enables sustaining a high tempo of operations that, when combined with responsive adaptive forces, will allow U.S. forces to "turn inside" the enemy's decision and action cycles; these real-time, mutually supporting adaptive capabilities will create unexpected opportunities because the opponent will never quite be in step. And maximizing individual initiative and enabling operations with a minimum of deliberate planning or preparation will allow U.S. forces to exploit opportunities created in this complex battlespace. Dominating the air and space medium enables our innovative operational concepts, protects our forces and functions, and denies our adversary the choice of actions and responses. This more integrated model, in return, offers a substantial *multiplier* effect—increased force effectiveness and operational adaptability from greater integration at the tactical echelons. It increases, on the other hand, the difficulty of creating "jointness" because it requires both integrated doctrine and operational integration at lower tactical echelons.[15]

In this concept, as opposed to the more traditional views of air and space power as the supporting force, air and space power is seen as the lethal element—indeed, the hammer and anvil analogy noted earlier is inverted, with ground forces (smaller, more agile, lighter, less tethered) now becoming the means to prepare the battlespace and shape the enemy force and its tactical configuration. This concept would maximize the ability of air power to be devastatingly lethal against massed forces by combining the air power with ground maneuver elements that force the opponent to concentrate or be defeated in detail by the integrated force.

Finally, we need revised concepts for how new space and air capabilities are integrated with other forces. One option is full integration through an expanded, enhanced cooperative engagement capability (CEC). These new concepts would use many of the same technologies demonstrated during the Gulf War—stealth, effective SEAD, proliferated space sensors, real-time airborne battle management—but they would be integrated into an operational concept for continuous air operations in order to dominate all operations in the battlespace. Changes in aerospace operations (for both air and space forces) from emphasis on a preplanned episodic/periodic model to one stressing real-time adaptivity, continuity, and persistence would necessitate significant alterations from historic air force and navy concepts of operations for air and space systems. But the new operational patterns that these would allow would provide a significant opportunity to develop a fully integrated joint operational approach that maximizes the potency of U.S. forces and enables the achievement of politically effective operational results without overwhelming force. Continuous, persistent air and space operations, instead of episodic appearances in the battlespace, would permit air forces to create an "air overwatch" capability to dominate all tactical operations beneath.

Such changes in the patterns for future aerospace operations stressing continuing "control" of the operating environment through integration of all elements of air and space power would be a dominant influence in effecting changes in ground and naval force operations. These changes, however, would not be brand-new but would build on evolving concepts and operational capabilities demonstrated during the Gulf War such as real-time tactical support from space, JSTARS, airborne warning and control systems (AWACS), F-15E loiter-mode operations, and missile defense, among others. Long-endurance UAVs, stealth, advanced ECMs, and continuous multiple coverage from small tactical satellites, among others, are all new technologies that can enable new missions such as overwatch and air constabulary functions. Together, these capabilities would lay the foundation for implementing the concept of "air occupation"—control of the "high ground"—that Gen. Ronald Fogleman has enunciated. Without this ability to provide truly responsive real-time air and space power, forces relying on air power can react only to an opponent's initiatives and actions; with real-time responsiveness underwritten through these concepts, U.S. forces can operate inside the enemy's decision and action cycle, truly dominating the entire spectrum of combat operations. Moreover, these capabilities would provide a seamless integrated operation across the strategic, operational, and tactical levels—capable of converting even local breakthroughs into decisive operational results.

Perhaps more than anything else, both of these new operational concepts and new objectives for air and space power will demand new thinking and new emphases from air and space power proponents. Asking to improve the old is not wrong, but it misses the opportunity, not often granted, to fundamentally reshape air and space power in keeping with new circumstances, new capabilities, and new priorities. If we are, indeed, moving toward an information age, one in which C^4ISR will be the core capability for all U.S. forces, then air and space power advocates should seize this moment and build their future concept around this, rather than around delivery of lethal mass.

Assessing Air and Space Power for Strategic Relevance

We appear to be at a turning point in history. Although we can optimize air and space power, and the entire military establishment, based on existing tactical and operational problems, that would seem to misapprehend the real national security problem that we face. We need to be able to maintain our military strength and the flexibility to meet whatever appears "on the other side"—whether "near-peers," peers, or other threats not yet characterized. This demands that we not forfeit the support of the American people or don a straitjacket by taking too focused an approach in the interim.

In attempting to balance different national security priorities that compete for current resources, we are really attempting to balance the risk among problems that have different time horizons. In essence, the metric for strategic choices in the near term must construct a risk and time-discounted investment portfolio that reaches out into the future priorities because even current choices will have long-term consequences. The proposed framework may provide one way of assessing these choices.

With the demise of the Soviet Union and the day-to-day nuclear standoff between the two superpowers over, unless Russia significantly reverses course and resumes a hostile posture, the United States no longer faces a direct threat to its national survival in the near term. What we do confront in this period are a wide range of threats to international order. These may require U.S. attention and intervention, but they should not be used to shape U.S. forces precisely because of the long-term consequences of U.S. investment decisions. Over the medium term, the period that the major modernization programs are designed to address, the United States does not appear to face a challenge to its military superiority, either technological or quantitative. During this period, no potential opponent has sufficient strength to threaten our overwhelming advantages if we show proper caution. Given the immense technological advantages we now possess, coupled with our large, well-trained, and well-equipped forces, we do have the capacity—with timely adjustments to our planned forces—to match any nonpeer challenger that might arise.

Over the long term, however, though it is currently impossible to predict with any certainty who a challenger may be and it is infeasible to constrain the character of the challenge we may face, it would be dangerous to assume that over a 20-year period no challenger could pose a threat to our vital strategic interests. Neither concepts formed in an era of overarching nuclear forces and concentrated combat across the inter-German border nor those developed to fight regional hegemonies under less demanding circumstances appear to be appropriate for addressing potential new strategic challengers. A new American way of war provides the basis for a new conception of air and space power as well as a foundation for assessing the merits. Is there any other choice?

Observations

The approach outlined in this essay does not differ from current policy or force structure or technology; rather, it proposes that a different strategy and doctrine be adopted for how air and space power is employed—and, therefore, for how the merits of air and space power are assessed. This strategy builds on air and space capabilities as the decisive component of a truly integrated joint force and thus differs from approaches that view air power either as an independent element or as merely a supporting arm. To begin the transformation from air and space power as a supporting force to a force capable of wielding decisive power worldwide, it is critical that investments be made now, along with constructing the doctrine and fleshing out the conceptual basis for this new strategy. This transition cannot be made with piecemeal or grudging support from the U.S. Air Force itself while it is attempting, in a period of declining defense resources, to promulgate the doctrinal sources of traditional air power or maintain force strength in the classic mode—the platforms (or "rubber on the ramp").

Moreover, it is unlikely that the transformation to a new American way of war can be accomplished if the deck chairs are merely rearranged and the traditional budget slices allocated among the armed services are retained. No service will voluntarily surrender claims on resources in a period of declining defense budgets. Therefore, the Department of Defense, assisted by Congress, will have to reshuffle the budget deck to produce a new deal that reallocates scarce resources in a manner that is consistent with challenges of the future, not those of the past. Although air and space power proponents have a strong case for receiving a larger share of defense resources, it is important that they construct this case in explicit recognition of the necessary contributions of other force elements to an effective integrated operational concept—not on a foundation of air and space power as the independently decisive force. Air and space power may be the hammer, but the anvil is essential. The figure of merit should be overall force effectiveness, not the number of tactical air wings retained or the increase in budget share.

Within the air power slice, budget decisions are, to a large degree, not as much a question of investment choices among advanced air and space platforms as they are of the balance between platforms and leverage-enabling elements. While arguments rage over the increasingly high proportion of "tail to tooth," a driver for this shift lies in the dramatically increased lethality of modern platforms equipped with brilliant weapons, both underwritten by advanced C^4ISR systems. To the extent that we can maintain military effectiveness while putting forward fewer forces to face direct combat hazards, an increased "tail to tooth ratio" is a sign of an improved and a more sustainable military posture. Thus, increased spending on the critical C^4ISR backbone, on focused logistics, and especially on large-scale procurement of high-leverage advanced precision weapons would help to implement the transition to more effective concepts for combat operations. We should expand the current program of advanced concept and technology demonstrations (ACTDs) to provide continual doctrinal reassessment and institutionalize permanent revolution in concepts of operation through innovative approaches to concept development and intensive field experimentation. The greatest danger we run in being overtaken or successfully challenged militarily is by standing still and giving potential opponents a static operational target.

It would also be a mistake to try to use the defense planning processes we adopted during our confrontation with the Soviet Union to develop and assess new strategic approaches to evolving geostrategic challenges—the character of these challenges is fundamentally different, and our role has significantly altered. Attempts to find or define suitable threats so that we can continue to do "threat-based planning" are therefore doomed to run afoul of the American public, which has become increasingly wary of manufactured dangers; moreover, there is some danger that reactions to "being fingered" by potential challengers may produce self-fulfilling prophesies. Restructuring the planning process and building on a capabilities-based approach that defines broad operational needs would be a useful first step.

A second step involves addressing the problematic state of our foundation for assessing combat effectiveness and future force needs. Current models and simulations that are used to appraise future weapons systems, especially those for air and space, appear irrelevant to the future operational concepts and combat conditions. At best, they appear to yield poor results even for existing systems and concepts of operation; at worst, they seem to produce orthogonal results and dysfunctional solutions at variance with the best evidence from actual combat experience. The inability of existing combat models to produce results that replicate recent historical experience suggests that the basic relationships and physics of the combat processes embedded in these models are faulty.

Nowhere are the models more vulnerable to challenge than in their treatment of the critical human element in combat—including the role of commanders' decision processes. Human factors may have as much to do with a unit's military effectiveness, especially in future high-tempo distributed combat, as the technical performance of the weapons systems. Perhaps the least noticed element of the U.S. advantage during the Gulf War was the superbly honed edge of U.S. forces that had been produced by intensive training, extensive exercises, and sophisticated simulation. No other nation maintains the training, exercise, and simulation base needed to conduct realistic large-scale exercises that allow the United States to test and improve not only weapons, but especially doctrine and operational concepts under field conditions. Yet we currently have no basis on which to confidently assess the magnitude of these factors nor the relative share of the investment resources that they should fairly receive.

Finally, the increasing proliferation of capabilities for weapons of mass destruction, the diffusion of advanced conventional weapons to third parties, and the clear dangers to deployed forces from terrorists all suggest that future operational concepts need to consider security and survivability (i.e., force protection) as an inherent element of operational concept design, not just an overlay or appliqué. Future operations, whether contingency responses to crises or acts of overt aggression, on the one hand, or long-term presence deployments, on the other, must be conducted as expeditionary warfare in mind-set as well. The past two decade's events, from the bombing of the Beirut marine barracks to Khobar Towers, should convince us of the need to develop operational concepts, including those for the entire rear area and logistics stream, suited to hostile, nonpermissive environments. Current air operations depend on a complex choreography, and tactical operations, especially those conducted from in-theater bases, are particularly vulnerable to disruption of the careful sequencing of essential tasks and elements. Placing another ring of concertina or assigning more guards with night vision devices is no substitute for an inherently well-designed operational concept and posture. Just because we will increasingly be responding to contingencies that are unpredictable does not mean that our operational posture must be ad hoc as well.

Notes

1. Russell F. Weigley, *The American Way of War: A History of United States Military Strategy and Policy* (Bloomington: Indiana University Press, 1973).

2. Future combat environments will likely be *nonlinear* in two ways. First, in the military sense of the term, there will be fewer situations with continuous lines of troops and defined forward edges of the battlefield (FEBAs) and more dispersed operations by discrete units operating over sparse battlefields. Second, in the mathematical sense of the term, outcomes will increasingly appear to disproportionate to the force inputs, and, therefore, classic rules of thumb will need to be discarded. Such combat outcomes will often be unexpected (emergent), require real-time decisions and actions to exploit, and serve to tightly couple tactical, operational, and strategic levels of combat.

3. "The dogmas of the quiet past are inadequate to the stormy present. . . . As our case is new, so we must think anew. We must disenthrall ourselves, and then we shall save our country." Abraham Lincoln, Second Annual Message to Congress, December 1, 1862.

4. Weigley, *American Way of War.*

5. Clearly, some air power leaders did believe that air attacks on the enemy homeland could bring victory and obviate the need for invasions, but the strategic plans presented to and endorsed by the national leadership did not adopt this course.

6. In the case of Japan, air power was combined with an extremely effective naval blockade by submarine.

7. See Department of Defense, Title V Report to Congress: *Conduct of the Persian Gulf War.* Not many in the U.S. Air Force would support this line of argument as put forward by Col. John Warden.

8. See, e.g., Christopher J. Bowie et al., *The New Calculus: Analyzing Airpower's Role in Joint Theater Campaigns* (Santa Monica, Calif.: RAND, 1993).

9. The Gulf War Airpower Survey (GWAPS) provides substantial detail for the adverse impact of bad weather on air operations. On Days 2, 3, and 7 of the crucial first week of air operations, more than 60 percent of F-

117 weapons were weather-related no-drops, cancels, or misses—about 35 percent of total F-117 missions in Week 1. This situation mirrors that of the 8th Air Force during the Combined Bomber Offensive against Germany when its commander stated that weather had been a bigger obstacle than the German Air Force.

10. This system is based on Thomas Malone's tripartite taxonomy of the effects of technological change, described in Thomas W. Malone and John F. Rockart, "Computers, Networks, and the Corporation," *Scientific American* 265, no. 3 (September 1991): 128–136

11. GPS was strongly resisted by many in the air force community who feared that it might be unreliable and subject to countermeasures and outages, or lead to loss of tactical independence.

12. This notion of "setting the terms" for subsequent combat operations is similar to the Soviet doctrinal notion of "initiation, course, and outcome"—a related sequence of activities leading to victory.

13. Although coverage from overhead assets can be adjusted, it is extraordinarily time-consuming and expensive, especially in opportunity costs. Therefore, in most cases greater reliance on alternatives for these pre-conflict periods is essential.

14. A critical problem is that acquisition of increasingly longer-range tactical ballistic missiles will place fixed facilities at great risk. Unless these facilities can be protected by TMD assets deployed prior to the crisis (whether host country or U.S. operated) or by offshore naval-based TMD, the enemy can make it extremely difficult to establish a survivable lodgement as the very airheads needed for deployment of the airlift-intensive ground-based TMD are themselves subject to attack and interdiction. Thus, it may take a considerable time—until the threat is sufficiently under control—to put very high-value, soft assets such as airborne warning and control systems (AWACS), JSTARS, and HALO/HALE UAVs at risk at fixed ground sites. Therefore, it is crucial to exercise these capabilities from an expedient and sustainable expeditionary posture.

15. See, e.g., Gen. John Sheehan, "Next Steps," *Joint Forces Quarterly* (Autumn 1996): 41 ff.

3

Air and Space Superiority

Richard P. Hallion and Michael Irish

Declining resources and changes in the global political environment are forcing a reexamination of U.S. military strategy and structure. This reexamination, presently framed by the Quadrennial Defense Review and the National Defense Panel, has the potential to leverage new capabilities with equally powerful operational concepts, providing powerful enhancements to our national security in the twenty-first century.

Air and space superiority offers the national political leadership the freedom to engage globally at any time and in any place—in sum, the freedom to exercise national prerogatives. It is the critical, synergistic enabler for all forms of military power, ranging from other air power projection forces to ground and naval forces. With air and space superiority, the United States can use the full range of its air power forces to maximum effect, including ones that might otherwise be vulnerable to an enemy contesting control of the skies. Air and space superiority reduces losses and markedly increases the effectiveness of friendly military operations. Overall, control of air and space provides both the freedom to attack and—in many ways, more important—freedom from effective counterattack. American forces have enjoyed the latter advantage since mid-1943, overwhelmingly so since the Korean War.

In World War II, Anglo-American air power so dominated the battlefield that Nazi commanders, including Field Marshal Erwin Rommel, complained of being fixed in place, denied the ability to maneuver, and forced to endure battlefield paralysis that prevented them from fulfilling their strategic and tactical plans. By war's end, the vast majority of enemy casualties and materiél losses were being inflicted by air attack and artillery, not by infantry weapons, beginning a trend that has continued to the present day, through the Korean, Vietnam, and Gulf Wars.

Air and space power's primary merit is that it furnishes the fastest and most responsive means of projecting power, presence, and influence. In the era of predominant land warfare, typified by the Roman model of warfare, forces moving on foot could control 30 percent of the earth's surface at a mobility rate of approximately 1½ miles per hour. In the era of predominant sea power, typified by the British Empire in the early twentieth century, forces afloat could control 70 percent of the earth's surface and coastal regions by moving across the ocean at a mobility rate of approximately 20 miles per hour. In the era of predominant air power, forces operating through the air can control virtually 100 percent of the earth's surface at a mobility rate in excess of 500 miles per hour. Forces using the space medium can traverse the entire globe at 18,500 mile per hour, reaching antipodal points in approximately 45 minutes.

Because of its growing air power advantage, since 1945, the principal means whereby the United States has signaled its interest and resolve in crisis situations has shifted from sending large land and sea contingents to sending air power forces, typically deploying aircraft carrier battle groups, air expeditionary forces, or, more characteristically since the 1980s, airborne warning and control system (AWACS) surveillance aircraft, into crisis regions. The vital and synergistic partnership of *air and space superiority* and *precision engagement* has redefined crisis response, assisted greatly by *global air mobility.*

Air and Space Superiority: The Critical Enabler

Critical to our nation's ability to prevent or respond to crises and contain conflicts are capabilities, technologies, and systems that enable a rapid global application of power with minimum cost and risk. In *Joint Vision 2010*, the chairman of the Joint Chiefs of Staff outlines four operational concepts required for military dominance into the next century. They are *dominant maneuver, precision engagement, focused logistics,* and *full-dimensional protection. Air and space superiority* is the critical enabler for them.

A few years ago we witnessed in Iraq the fate of a country that lacked air and space superiority. Fortunately the United States had all the ingredients necessary to overcome Iraq's air power and air defenses. This achievement required F-117 stealth attack fighters to knock out air defense control centers while F-15 air-to-air fighters patrolled the skies and kept Iraq's fighter aircraft in their supposedly impregnable concrete shelters—which our laser-guided bombs proceeded to penetrate one after another, forcing surviving Iraqi aircraft to flee to Iran. Unable to fly over coalition forces (let alone attack them from the air), Iraq lay blind to the air and ground offensives gathering against it, while American aircraft and satellites exposed Iraq's plight to constant observation. Since then our aging fighters have enforced peace by patrolling Iraq and Bosnia day in and day out. Anyone seeking to challenge the United States can deduce that they will succeed only if American air and space superiority can be overcome.

Air power has evolved beyond being a supporting arm of the surface forces and has matured to achieve many of the capabilities dreamed of by early air power advocates. Although some circumstances might allow its independent use, the true value of air and space power as an instrument of national power is its capability to be a leading force, making all of the component forces more effective in the coherent joint campaign. Prior to Operation Desert Storm, predictions were made about casualties in the tens of thousands and a war that would last more than a year. Neither came true because we fought much of the war with air and space power in a manner that Iraq had not anticipated. Instead of directly confronting and attempting to displace Saddam Hussein's troops from their dug-in positions with tanks and troops, we asymmetrically applied air and space power to cripple their air defense systems, command and control, and infrastructure. With national space assets we targeted cruise missiles and precision-guided munitions that F-117s delivered with impunity into the heart of Baghdad.

As many commentators noted, Desert Storm was a signal that land-centric doctrine, strategy, and training had not only lost their preeminence but also become a recipe for defeat by modern joint forces. For Russia in particular, the Gulf War marked a watershed, as the demise of its former client state, Iraq, called into question over a half century of Soviet military thought. Ironically, two months before Saddam Hussein marched into Kuwait, the commandant of the Military Academy of the General Staff, USSR, Gen.-Col. Igor Rodionov, commented in a lecture to the U.S. Naval War College, "For hundreds of years, victory in war was achieved through the seizure of territory. That stereotype no longer applies." In September 1990, barely one month into Operation Desert Shield, his colleague Gen.-Maj.V.I. Slipchenko argued in a lecture to the U.S. National Defense University that "the traditional role of conventional armed forces equipped with infantry, tanks and artillery is virtually eliminated. . . . The revolutionary change in military art leading to the 'future war' concept . . . is beginning. The leaders of both our countries must deal with it." The lessons of history are clear: the next generation of military thought must focus on the exploitation of air and space power. The United States, which invented the airplane, should not fall behind as it did before World War I and again before World War II and again before Sputnik.

Though the predictable bipolar threat under which the West lived for over 40 years has ended, the confluence of two dangerous trends has accelerated the need for a radical advance in American air superiority capabilities. The first of these is the large number of modification programs and the proliferation of advanced sensor and missile systems that are being added to older-generation fighter aircraft, coupled with new generations of fighters such as the Su-35, Gripen, Rafale, and Eurofighter. The second is the understandable—and necessary—drawdown of American military forces. It is clear that, in its modernization programs for future fighter and strike aircraft, the United States must not ignore the need to maintain and even increase its capabilities to control the air and space domains.

What Distinguishes Air and Space Superiority

Air and space superiority is the control of the air and space in a theater of operations to a degree that enables the free prosecution of offensive operations and insulation in defensive operations in all battlespace dimensions—air, sea, land, and space.

Air and space superiority offers the following advantages:

- provides for all services the freedom to *operate*, including freedom *from attack* and freedom *to attack*.

- puts the *enemy in a position of severe disadvantage* or holds him hostage.

- ensures *freedom of movement* on the ground, at sea, in the air, and in space. This means having the ability to deploy forces, and to move people and cargo for wartime or peacetime operations, as well as protecting our nation, our allies, and our forces based overseas from air and missile threats.

- preserves *unencumbered use of space assets* to exploit advances in space and information technologies.

Air and space superiority requires the capability to conduct the following missions:

- *Offensive counter air:* Missions that take the initiative to destroy the enemy's ability to operate in the air environment by attacking enemy systems—both in the air and on the ground.
- *Defensive counter air:* Missions that protect against attack from enemy systems that operate in the atmosphere—systems that include aircraft, cruise missiles, theater ballistic missiles, and surface-to-air missiles (SAMs).
- *Offensive counter space:* Missions conducted against an adversary's space-based systems.
- *Defensive counter space:* Missions that defend against systems operating in space, including missions that protect against cruise and theater missile defense—e.g., ballistic missile defense.

These fundamentals are essential to successful power projection and provide a solid framework for strategic planning. Total command of the air and space means that you control the ultimate "high ground" of warfare, and all operations are possible. Conversely, if the enemy controls or impedes the air and space dimension, the simplest land maneuver, or resupply effort, is in jeopardy. This was true in World War I, the Spanish Civil War, World War II, Korea, Vietnam, and the Gulf.

Air and space superiority offers the national military leadership the freedom to engage globally, at the time and place of its choosing, while denying the enemy freedom of operation. Because it provides forces of all services the freedom to operate, air and space superiority increasingly has become the aegis and critical underpinning for virtually all military operations.

Air and space superiority requires the robust application of our best technology for platforms, weapons, and systems. It cannot be done "on the cheap." It demands a system of systems, not a one-size-fits-all solution. It is enabled by sophisticated platforms, managed with sophisticated command and control, and leveraged by operational art and integration. Air and space superiority leads the way to control of the terrestrial battlefield: tanks cannot hide or maneuver under the watchful eye of the Joint Surveillance and Target Attack Radar System (JSTARS); enemy aircraft cannot fly at will when confronted with F-22s and

AWACS; and even the best-defended nations cannot protect their command and control facilities against attack when faced with the stealth and precision bombing of the F-117, the F-22, or the B-2.

Desert Storm taught a few simple, yet powerful lessons. Air superiority can turn a massively armed aggressor into an ineffective force—leading to a relatively quick defeat. Air dominance can dramatically limit U.S. and coalition casualties. Presently, the United States has the ability to dominate many of the mission areas central to air and space control. From counter air to strategic attack; from air refueling to air resupply, and from command and control to information warfare, we have distinct advantages. Though space control is a relatively new area in terms of policy, doctrine, and strategy, it is directly linked to the success of all military operations—whether through intelligence, communications, early warning, navigation, or targeting. Air and space superiority will become increasingly important to effective joint and coalition operations.

The Air Superiority Challenge

Whereas the concept of and need for air superiority is largely unchanged, how it is achieved *is* changing. Air superiority has always been a required "state" rather than an optional "mission" because it enables all operations undertaken by the joint force commander. This is reflected in *Joint Vision 2010*. What has changed is that it can now be achieved by a variety of means, far beyond traditional notions of dogfighting or bombing.

Military operations demand precision for both lethal and nonlethal applications of force. We require precise, high-fidelity imagery and information to identify targets. As we increasingly rely on precision munitions, the dependence on space-based sensors will increase. Desert Storm demonstrated the capabilities of these technologies, however immature. Unparalleled situational awareness was provided by real-time information from platforms such as the RC-135, AWACS, JSTARS, and the U-2 in addition to national space assets. These air and space tools provided the commander with unparalleled and unprecedented clarity as to the disposition of the battlespace.

Due to the increasing lethality associated with weapons delivered from and through the air, air superiority is now recognized as essential to success on the battlefield, both in major theater warfare and in more limited contingencies. Without air superiority, Saddam Hussein was unable to detect allied force movements or to disrupt them with his air and ground forces. As a result, his only options were to try to split the coalition or provoke an attack through unconventional means, such as Scud missile attacks, or to await allied attack operations at the time and choosing of the coalition, while his war-fighting capability was continuously eroded by allied bombing.

Air superiority is a prerequisite for both large and small contingencies, including peacekeeping and even humanitarian operations, as recent experience in Bosnia and Rwanda has indicated. In Bosnia, F-16 fighters shot down four Serbian aircraft violating no-fly restrictions, essentially bringing hostile (and potentially hostile) air operations to an end. Then, in the NATO air campaign, the devastating effect of pinpoint precision attack against key Serbian targets by allied aircraft exploiting the targeting and attack options enabled by air superiority was the key reason why the Serbs chose to negotiate rather than prolong that civil war.

Achieving air superiority almost always entails significant effort and risk, particularly if one is operating against a more numerous foe. The dazzling success of Operation Desert Storm may, in fact, have obscured the fact that even this operation demanded a high level of planning and execution. Air superiority over Iraq required the full gamut of defensive and offensive air operations, missile defense operations, focused suppression of enemy air defenses, and effective information operations to assure control of the air.

Offensive Counter Air

Missions that take the initiative to destroy the enemy's ability to operate in the air environment by attacking enemy systems, including all aircraft, unmanned aerial vehicles (UAVs), and missiles, are considered offensive counter air. Used as the first step in gaining air superiority, offensive counter air operations enable attacks against enemy centers of gravity. This effort is more than just aircraft versus aircraft. Attacks on command and control, intelligence, air defense, airfields, fuel and maintenance storage, and other supporting facilities are required so that an enemy air force cannot challenge the United States in the air.

In Desert Storm, the suppression of enemy air defenses (SEAD) was the key phase of the fight for air superiority. Long-range artillery, cruise missiles, stealth aircraft with precision weapons, attack helicopters, and dedicated suppression aircraft carrying anti-radiation missiles or standoff precision weapons were used to punch holes in the enemy's air defense net, providing avenues through which other strike aircraft could fly to attack airfields and other key targets deep inside Iraq. These strike aircraft were accompanied by air superiority aircraft whose job it was to ensure than any airborne enemy fighters or high-value command and control assets were destroyed or rendered ineffective.

These operations were orchestrated by a joint forces air component commander (JFACC) who used all of the assets at his disposal to render ineffective Iraq's ability to affect air operations. The integrated application of force by the JFACC eliminated threats close to the border, destroyed key targets deep inside of Iraq, and orchestrated effective information operations designed to enhance the joint force commander's view of the theater and deny that same capability to Iraq. This same use of force would be applied to any contingency in the future.

Defensive Counter Air

Defensive counter air missions protect against attack from enemy systems that operate in the atmosphere. Similarly, defensive counter air and space missions reactively protect against attack from enemy systems that operate in or transit through the atmosphere or space. Counter air is an effect, not a means. All actions that prevent the enemy from exploiting air and space while allowing our forces to exploit air and space are counter air. This includes defensive measures against ballistic and cruise missiles as well as manned and unmanned fixed-wing platforms.

Without defensive counter air operations the United States cannot successfully defend a position or prepare a counterattack. Failure in this arena also limits U.S. Army access to close air support, as these assets will themselves be fighting for their existence in the face of the enemy air threat. Without air superiority, it is impossible to protect U.S. troops from enemy air attack. Defensive counter air operations in the Gulf War included both airborne and surface systems working together to ensure that no enemy aircraft were able to attack friendly forces on the ground or in the air. Once the defensive job was assured during Desert Shield, systems such as the J-STARS surveillance and targeting platform, the AWACS airborne control system, and a broad range of manned and unmanned airborne reconnaissance systems were used to find, fix, and provide targeting data that would prove essential in the initial stages of the air campaign. This intelligence preparation of the battlefield enabled air and selected ground and sea forces to suppress enemy air defenses and eliminate key command and control nodes in the first waves of attack.

Space Superiority

The revolutionary advances in aeronautics in this century are currently being matched by the increasingly explosive growth of astronautics. As space technology advances, so does the nation's understanding of exploiting this new medium for military operations. Further, the traditional bipolar space rivalry between the former Soviet Union and the United States has, in the post-Cold War era, given way to a more complex space picture as a variety of large and small nations on virtually all the populated continents attempt to secure some form of a space future.

From the U.S. perspective, we have already fought our first space war. In the Gulf War, space-based assets furnished intelligence, weather, navigation, communication, warning, and cuing for coalition military operations. Space capabilities support an increasing percentage of forces. Traditionally, space has been used as a sanctuary from which satellites can provide information to commanders about enemy movements and force dispositions. Space capabilities are increasingly vital to U.S. national security interests and are linked to military operations on land, at sea, and in the air. Missions are likely to increasingly evolve from being supported from space to operating in space.

Space is part of the environment that the military has long operated in. Weapon systems transit space (intercontinental ballistic missiles [ICBMs] and theater ballistic missiles [TBMs]). Space provides direct situational awareness to ground forces (the global positioning system [GPS]). And space-based sensors are essential for strategic and tactical warning. Although space operations have been considered uniquely different from those in the air, on land, and at sea, in the long term the differentiation between air and space will diminish. Today, unhindered space access and utilization is not merely "nice to have," it is critical for all American military operations, from low-intensity to high-intensity conflict, and from missions of presence to humanitarian relief. Under these circumstances, space superiority becomes as necessary an attribute as air superiority.

Space superiority operations are those that provide freedom of action in space for friendly forces while denying it to an enemy. They include the broad aspects of protection of U.S. and U.S.-allied space systems and negation of enemy space systems. With the addition of space as the upper end of the aerospace spectrum, the same principles proven over the past 80 years of warfare apply: air and space is the key enabling medium for victory in conflict, one that transcends the land and sea layer that forms its only boundary. Controlling that medium is the critical enabler both to assure any nation's ability to project force globally and to deny that same capability to adversaries.

Whereas achieving air superiority is increasingly well understood, space superiority may take on a slightly different definition. Space superiority provides the ability to exploit space as a critical integrator and enables a range of operations. Unlike air superiority, however, space superiority can be achieved and maintained in peacetime. The critical issue is, can it be retained in time of conflict? Jammers and increasingly sophisticated information operations using technology that proliferates through such systems as the Internet can potentially impact our ability to retain space superiority in war. Commercial satellites and the widespread development of communications, navigation, and weather satellite systems by many nations ensure that the information and support generated by space systems will be available to even the poorest nations or threat groups. Destroying these systems to deny their availability to the enemy will have the effect of alienating nations or international commercial or treaty concerns that might not otherwise be involved in the conflict. Thus, assuring space superiority in battles of the future will require whole new technologies, concepts, and a diplomatic framework that protect U.S. and allied systems while selectively denying the capabilities of commercial and third nation systems to the enemy.

Emerging Threats to U.S. Air and Space Superiority: Investing for the Future

What is at stake is the future. Our national advantage attained through air and space superiority is slipping as next-generation fighter, satellite, missile, and information technologies proliferate internationally. Dramatic advances in next-generation fighters, both in avionics and weapons systems, point to *parity* with the 25-year-old F-15. Parity is further encouraged by the development of ever more sophisticated SAMs and air defense networks. The proliferation of long-range ballistic missiles poses yet another challenge to U.S. air and space superiority. Responding to this threat requires a range of both offensive and defensive measures to counter ballistic missile threats to U.S. forces, allies, and our territories. Adequate defense against ballistic missiles will inevitably involve a hitherto unknown degree of jointness and integration of air and space capabilities, both offensive and defensive. The anticipated widespread—indeed, global—sale of these aircraft and missile systems does not bode well for assuming that American air superiority is a given. Further, American superiority is far less certain as warfare shifts to information-intensive assets—assets that not only exploit or generate information but also are themselves critically dependent on it. Superiority will require a substantial investment in a number of air and space platforms and in a whole new generation of weapons. Supersonic flight without afterburners (or super-cruise), stealth and counter-stealth, UAVs, space-based offensive and defensive weapons, nonlethal technology, and information warfare will all shape the frontier of warfare where air and space superiority will be redefined.

Today, American air and space capabilities are being challenged and can be, at best, considered robust but slipping. The United States currently enjoys air and space superiority, but that superiority is not permanent, ordained, or guaranteed. Information technology is increasingly available to anybody willing to pay for it—including potential adversaries. Of greater concern is the exploitation, vulnerability, and control of technologies related to air superiority, overhead space systems, systems integration capabilities, and sophisticated communications. Many technologies are no longer the exclusive domain of the United States and other technologies are quickly evolving in the international commercial arena at a rate that may leapfrog U.S. or allied military investments. In one example, many of the more formidable aircraft that bid to threaten the F-15's control of the skies exist not on paper or in the data banks of some computers, but on flight lines today. When these prototypical designs—the Rafale, Gripen, Eurofighter, and Su-35, among others—enter operational service, the air superiority that this nation has possessed for the last quarter century will no longer be a given.

Additionally, the emerging regional strategies of anti-access (e.g., using combinations of intelligence, surveillance, and reconnaissance [ISR], mobile ballistic missiles, aircraft, cruise missiles, and chemical munitions) challenge current capabilities and freedom of maneuver. Individual focus on speed, or stealth, or precision will fail. What is required are systems that incorporate such qualities *in combination* to yield the necessary and sufficient effects to ensure that our air and space superiority capabilities remain secure.

Powerful technologies, especially information technologies, are proliferating and recognize no borders or boundaries. The global positioning system is an excellent example. Although this technology can make aircraft or automobile navigation simple and accurate, it can also be used for navigation of a ballistic missile, cruise missile, or unmanned aircraft packed with explosives. This technology is available virtually anywhere to anyone. The threat is that previously low-technology weapons such as the Scud missile can now be made into highly accurate precision-guided munitions for several hundred dollars.

Although commercial assets are presumed to be less robust than military and national assets, this is changing as commercial research and development (R&D) budgets are enhanced and increasingly approach the size and scope of nationally funded programs. We should remember that although the computer originally emerged from defense-related research, once it found its way into the commercial sector, technology cycle times, innovation, and cost all leapfrogged the federal government's expectations.

The Evolving Threat to Continued U.S. Air and Space Superiority

A variety of threats from airborne and ground-based systems, when coupled with the aging and nonstealthy nature of U.S. air superiority forces (the F-14, F-15, F-16, and F-18 fighter families, all of which date from the late 1960s or early 1970s) pose serious concerns for the future of American air superiority. The proliferation of information technology that can be harnessed to military systems is one important component of the threat. However, the United States will be faced with parity from several air-to-air and surface-to-air systems in the near future. The Navy F-14 Tomcat completed its maiden flight in 1970. America's premier air superiority fighter, the McDonnell-Douglas F-15 Eagle, first flew in 1972. Presently, the Eurofighter, French Rafale, Russian Su-35, and Swedish Gripen designs demonstrate the potential to approach and sometimes outperform the F-15 in maneuverability, range, or electronics. When fully mature, these aircraft will (in the most advantageous circumstances for the United States) be at parity with the F-15. In the worst circumstances, they will offer clear superiority over it, from basic flying qualities, to observability and detectability, to effectiveness of sensors and weapons. Thus the drawdown of U.S. forces, which has resulted in serious reductions in air superiority forces, is further exacerbated by potential technological inferiority. Moreover, because we have as a nation relied on superior fighter technology to permit older "legacy" aircraft systems to survive in combat, this change forecasts an era in which foreign developments will de facto eliminate America's ability to depend on these older legacy systems for a wide range of support and combat operations.

The air superiority threat is not limited, however, to just air-to-air threats. Significant developments in surface-to-air missiles pose equal challenges to nonstealthy or only semistealthy aircraft systems. For example, the SA-10 and the SA-12 are both proliferating rapidly throughout the world. In fact, by the year 2005, it is estimated that no less than 22 nations will have these state-of-the-art air defense systems. Because an operation can now be seriously constrained by the 25-year-old SA-6 or even the 40-year-old SA-2, the danger posed by such systems is both a warning and a goad to develop counters—particularly aircraft systems that can successfully confront, engage, and defeat such advanced weapons.

The Space and Theater Missile Defense Challenge

Cruise and ballistic missiles pose threats to American forces based both overseas and at home. U.S. forces face a dangerous threat from theater ballistic missiles, such as the Scud. This threat is proliferating rapidly throughout the world, and the accuracy of these previously crude devices can be leveraged through modern information technologies, such as the GPS. Attacking missiles while they are on the ground is the preferred option. But, if that is not possible, the next best opportunity is in the boost phase, when the infrared signature is significant. The least preferable option, known as the "catchers mitt," calls for interception as the missiles approach their targets. As we saw during Desert Storm, that is not always the preferred method, especially if the missiles are carrying chemical, biological, or nuclear weapons. The airborne laser (ABL) appears to be one of the few systems that promises to be effective against this threat.

Cruise and ballistic missile defense is vital to U.S. national security. Yet our defense capabilities against the cruise and ballistic missile threat are weak. We will need to ensure that we have the necessary defensive systems in place before the threat becomes painfully real. Indeed, at a time when ballistic missiles are proliferating and the technology for such weapons is increasingly available, we currently have no capability to defend ourselves from a ballistic missile attack launched against North America. Though the intelligence community assesses the likelihood of a ballistic missile attack as a "remote possibility," we should bear in mind that we have consistently, throughout history, misjudged the future. From Pearl Harbor to Kuwait, we have been surprised. We should pay heed to the possibility that weaker nations will use ballistic missiles asymmetrically against the United States if they feel threatened. Prudence requires that we begin the process of developing a national missile defense in case the threat emerges more quickly than anticipated. Further, our space-based assets for early attack warning are aging, and more accurate systems will be required in the future. Development and deployment of the full architecture of the space-based infrared system (SBIRS) as well as a robust ground-based surveillance system is critical to completing our integrated picture of the air and space enveloping North America.

Today there are over 500 satellites in space valued at over $100 billion—over 40 percent of those are U.S. satellites, about half of which are military. We critically depend on them both in peace and war. Although space capabilities are essential to maintaining and achieving key political, economic, and military objectives, they also immediately touch on commercial and international policy issues that have sweeping implications in all directions for policymakers. This has made development of clear offensive and defensive military policies somewhat problematic. We rely on space power to enhance our ability to carry out our terrestrial responsibilities; consequently, we must assure our access to space-derived information, while negating our enemies' exploitation of the medium. We need to better understand potential threats and take necessary measures to protect our space interests and investments.

These threats are not new and are accelerating, suggesting a time within the next decade when routine American military operations will be readily monitored by a variety of space-exploiting nations. More than a decade ago, a civilian satellite spotted, identified, and tracked a military aircraft. As with air superiority, the control of space-based ISR should not be presumed as a given. The military implications of proliferating commercial space technology are dramatic. Whether precision navigation, remote sensing, imagery, or handheld global communications, each holds the potential to challenge future military operations. Several nations are developing a robust space-based remote sensing capability. For several hundred dollars, high-resolution imagery is available. Some nations already maintain a mature commercial capability. In fact, imagery generated by French satellites is presently in use by the U.S. military. Even if not in real time, consider the advantages this imagery would have given Iraq and the subsequent threat that would have posed to our forces on the ground.

Presently, some 30 nations own or operate geosynchronous communications systems. The international commercial space market is still in its infancy but is growing rapidly. More than 50 percent of all international space centers are neither U.S.—nor Russian—controlled. More than 25 percent of all space launches are neither U.S. nor Russian. Many space-based assets are covered by international treaty and are not owned solely by American companies or consortiums. Consider the implications if we tried to contain the communications capability of China due to regional belligerence directed at Taiwan, and China is one of the treaty signatories to an international satellite agreement. Is denying use of international satellite systems an act of war, an economic blockade? Even worse, what do we do if our systems are destroyed? Can we afford to have our systems disabled? Does such disabling constitute an act of war, a space-age Pearl Harbor? What will the National Command Authority decide if the attack is against an international commercial asset and not on or above U.S. soil? These questions must be answered as space control continues to emerge as a prerequisite for success on the battlefield.

In short, we are faced with two major space challenges in the immediate future. One is to ensure access to space, and the other is to protect U.S. interests and investment in space. Increased reliance on space systems means improved capabilities, but also new vulnerabilities. We should accept that our space capabilities will be targets for enemy attack. The United States must be able to control the medium of space to assure our access, reduce our vulnerabilities, and deny the same to any adversary. We must dominate the military space dimension and integrate space forces into our overall war-fighting capabilities across the spectrum.

Air and Space Superiority: The Sine Qua Non for Future Military Success

Most nations, including the United States have consistently demonstrated the inability to accurately forecast long-range threats based on political intent or military objectives. We did not anticipate Iraq's push into Kuwait, just as we failed to predict either Nazi Germany's or militarist Japan's aggression, and we will likely make similar mistakes in the future. Because we cannot adequately predict which threats will emerge, and because we are unsure what future technologies will be available to others, we must base our national military strategy on the emerging capabilities of potential adversaries and then ensure that ours are superior.

Our nation requires capabilities that can manage the following situations. Future adversaries will know that they must reduce the degree of warning to the United States, if they are to accomplish their objectives. Future adversaries will invest in high-technology assets that can be used asymmetrically against us, such as ballistic and cruise missile technology, which impacts our ability to pre-position forces. Some may well threaten the use of weapons of mass destruction, including nuclear weapons. Future threats include

- *Regional and state-centered threats:* Several states are working to acquire weapons of mass destruction (WMD) and the ability to deliver them.

- *Transnational threats:* Terrorists, drug traders, and members of organized crime threaten American interests. In some cases, the technology available to these groups, when leveraged by modern information technology, provides them with power similar to that of some state actors.

- *Threats from WMD:* Weapons of mass destruction pose an even greater danger in the post-Cold War era. Both ballistic missile technology and various WMD technologies can be acquired by the dedicated organization. Previously crude delivery systems can be greatly enhanced through commercially available information and navigation systems.

The degree of coherence in our response to these threats may well define our success or failure. All the services, with a broad range of capabilities, will be required to effectively prosecute joint warfare. Air and space power can be used to assist, deter, compel, defend, or destroy. But regardless of the nature of our response, air and space superiority is often an assumed condition for the United States. However, our air and space superiority fabric is beginning to fray.

Future air and space superiority is jeopardized by three trends. First, in the current budgetary environment, cost cutting tends to follow a bureaucratic solution. This means that all military programs get their percentage of the cut, as opposed to a strategy-based solution. Consider the implications of a 10 percent decrease in sunset forces versus a 10 percent cut in a program that pushes the edge of future warfare. Second, the U.S. military has enjoyed its role as the premier air and space power so long that the nation has become overconfident of its ability to continue to dominate the competition—in the absence of significant investments in future air superiority systems. As other nations become ever more successful in the commercial aircraft industry, they find themselves more able to compete in the military arena. In conjunction with the proliferation of information systems and software, this is a critical threat. One of these nations with little investment in sunset systems may find it can invent the right new systems to neutralize the standing U.S. advantage. Finally, distracted by the need to conduct day-to-day business, funding current operations replaces funding for the weapons of the second and third decade of the twenty-first century.

The capacity to deploy rapidly and engage over heavily defended enemy territory, to preempt a potential air opponent, and to achieve first look, first shot, and first kill in combat paves the way for all other forces in most conflicts, including a major theater war. We must have the ability to penetrate heavy defenses and hit key targets on the ground. The theater commander in chief and the National Command Authority demand flexibility from all assets employed in a theater.

As the number of space-faring nations grows, the need for continued American space superiority will accelerate. In order to maintain superiority we must develop robust capabilities for control of space; continue to use viable nonmilitary space capabilities, services, and products when they are beneficial and cost-effective; and fully integrate space capabilities into air, land, and sea systems. Cruise and ballistic missile defense is vital to U.S. national security. Yet our defense capabilities against the cruise and ballistic missile threat are weak. We need to ensure that we have the necessary defensive systems in place before the threat becomes painfully real.

Today the currency of advanced air superiority demands first look, first shot, and first kill—attributes achievable only through all-aspect stealth, supercruise, advanced sensor fusion, great maneuverability, and integrated avionics that can communicate with space platforms and other communication assets. Without it, all joint operations will risk potential catastrophe. All-aspect stealth is a genuine military aerospace revolution that has so far exceeded the abilities of potential aggressors to develop or exploit similar capabilities or to field counters to it. It is, truly, the new edge in America's new aerospace way of war. This advantage must not be squandered.

4

Global Attack and Precision Strike

Jeffry A. Jackson

The merits of global attack and precision strike must be viewed within the emerging strategic environment. The U.S. requirement to defend its worldwide commitments and interests in the future places enormous emphasis on this capability. We may continue to rely on forward presence for influence, but the retrenchment of our forces to the continental United States (CONUS), the closing of overseas bases, the reluctance of states to grant access to expeditionary forces, and the increasing range of enemy weapon systems coupled with their improving intelligence, surveillance, reconnaissance (ISR) capabilities means that we must place a greater reliance on a long-range attack capability to exercise U.S. influence and to respond to both planned-for and unforeseen contingencies. In addition, the nature of precision strike capabilities has had a fundamental impact on the ways and means of force employment, especially for air and space power.

The U.S. reluctance to preemptively attack in crisis situations may grant our future adversaries the advantage of the initiative and therefore requires a U.S. ability to regain that initiative to accomplish our objectives. That ability may be derived from strikes at global ranges on short notice with devastating intensity and accuracy. The capability for such a response may deter the potential adversary and thus promote defense against a range of enemy capabilities. Global attack and precision strike are thus key to the U.S. national security posture.

As force posture and defense budgets are unlikely to grow, the United States will continue to rely heavily on advanced technology. The synergistic impact on the combined arms effort of air and space power's global attack and precision strike capabilities can allow the nation affordable investments to meet its global obligations. Allied contributions to the defense of mutual interests will remain important, but they are likely to lag behind the United States technically in the realm of air and space power.

However, we can share these technologies appropriately, thereby encouraging our allies to contribute important capabilities that we are unable or unwilling to provide (e.g., host nation support and manpower). Additionally, we may ask our allies to invest in complementary command, control, communications, computers, intelligence, surveillance, and reconnaissance (C^4ISR) capabilities to enable effective and interoperable combined force employment. Yet the U.S. capacity to strike around the world with precise effects should remain a fundamental ability of our air and space power.

Scope: Global Attack, and Precision Strike Defined

Global attack is the combined use of air and space assets to locate and strike targets anywhere on the earth. Although land and naval forces may be incorporated as part of a global attack campaign, both are range-limited and do not constitute a global attack capability in and of themselves. Air and space forces can overcome the range limitation through the inherent capability of a weapon or system, or through the range extenders of aerial refueling, global mobility, or forward basing.

For years, the intercontinental bomber and missile nuclear retaliatory forces constituted this country's only true global attack capabilities. Delivery accuracy was technologically limited and, with nuclear weapons effects, considered unnecessary. Countervalue targeting was the norm; the superpowers held each other's populations hostage to nuclear annihilation. As technologies improved, the delivery systems became sufficiently accurate to allow counterforce targeting to inflict unacceptable losses on the adversary's military capacity. The global strike capability of long-range bombers has increasingly shifted to conventional weapons, notably, precision-guided munitions.

Another means of global attack with conventional weapons is the Air Expeditionary Force (AEF). The AEF concept has been exercised in the Middle East to demonstrate a power projection capability deployed from the United States. Tailored to meet the joint force commander's needs, the AEF can launch and be ready to fight in less than three days for both lethal and nonlethal applications. The degree to which resources can be pre-positioned can improve the AEF's engagement and sustainment capabilities. The AEF allows the United States to establish a military presence and capability beyond the placement restrictions of the aircraft carrier. U.S. Air Force leaders point out that the

AEF is not a replacement for carriers, but rather a timely complement to give more options to decision-makers. The fast-moving, self-contained forces of the AEF place a potent capability into the crisis region to exert influence, deter aggression, or prosecute retribution.

Precision strike is the capability to attack targets with the exactness and intensity required to achieve the desired military effect with minimum collateral damage and a relative economy of force. Several conditions must be achieved for this level of accuracy and effect. The precision weapon requires accuracy in target recognition and identification, weapon aiming, weapon guidance, and terminal effects. Sensor and guidance options include electro-optical, infrared, radar, laser, anti-radiation, navigational (inertial, terrain-following radar [TFR], and global positioning system [GPS]) technologies. Precision strike capabilities have been adapted to both bombs and missiles (including air- and ground-launched and short- and long-range).

A precision strike regime, combining the properties of global attack and precision strike, would enable the United States to capitalize on its technological advantages in sensor systems; information systems; delivery platforms; accurate, precise, and brilliant munitions; and communications/feedback systems. Such a regime could lead to a dramatically different way of conducting future warfare where time, intensity, and precision are leveraged as unique sources of strategic advantage. These capabilities have a synergistic effect, making the rest of the combined arms team more potent.

The Merits of Global Attack and Precision Strike

The union of global attack and precision strike capabilities, combined with information superiority, has blurred the traditional distinctions between strategic, operational, and tactical operations. This was organizationally recognized to a degree by the U.S. Air Force (USAF) with the dissolution of Strategic and Tactical Air Commands and the creation of Air Combat Command. Consider the following Department of Defense definitions from the Joint Publication (JP) 1-02:[1]

- *Strategic level of war:* "The level of war at which a nation, often as a member of a group of nations, determines national or multinational (alliance or coalition) security objectives and guidance, and develops and uses national resources to accomplish these objectives. Activities at this level establish national and multinational military objectives; sequence initiatives; define limits and assess risks for the use of military and other instruments of national power; develop global plans or theater war plans to achieve these objectives; and provide military forces and other capabilities in accordance with strategic plans."

- *Operational level of war:* "The level of war at which campaigns and major operations are planned, conducted, and sustained to accomplish strategic objectives within theaters or areas of operations. Activities at this level link tactics and strategy by establishing operational objectives needed to accomplish the strategic objectives, sequencing events to achieve the operational objectives, initiating actions, and applying resources to bring about and sustain these events. These activities imply a broader dimension of time or space than do tactics; they ensure the logistic and administrative support of tactical forces, and provide the means by which tactical successes are exploited to achieve strategic objectives."

- *Tactical level of war:* "The level of war at which battles and engagements are planned and executed to accomplish military objectives assigned to tactical units or task forces. Activities at this level focus on the ordered arrangement and maneuver of combat elements in relation to each other and to the enemy to achieve combat objectives."

A problem on the modern battlefield is that what were once considered "strategic forces" can now be applied to theater and tactical situations. The boundaries between battles, campaigns, and theaters have become less and less definable because of the capacity for dominant battlespace knowledge and employment of forces across boundaries to shape that battlespace. Hence, the joint force commander might employ a "strategic" system to accomplish "tactical" objectives because of his ability to see the battlespace to that level and exercise the real-time command and control to assure mission success. The key then becomes how to use the assembled array of assets where they will have the greatest impact on achieving overall objectives, not in trying to distinguish or delineate between strategic, operational, or tactical missions or effects.

Although clear distinctions between strategic, operational, and tactical are less possible, it is nonetheless useful to examine the effect of global attack and precision strike capabilities at various levels of conflict. Whereas our future adversaries may meet us on conventional battlefields, the proliferation of weapons of mass destruction (WMD) demands that U.S. aerospace forces retain the capabilities required for robust nuclear deterrence across a range of contingencies and adversaries, including the demonstrated capability and will to invoke nuclear threats to deter nuclear, biological, and chemical (NBC) attacks in a major theater war.

One of the merits of global attack and precision strike at this level is that deterrence may be achieved with conventional weapons because of the ability to produce massed effects, precisely applied, without resorting to mass destruction. By maintaining the enormous U.S. advantage in global reach and power projection capabilities over any rival, the threat posed by our conventional forces will become more credible, and the possession and threatened use of nuclear weapons by any adversary, or in international relations generally, can be diminished to our advantage.

From World War II through the Gulf War to the present, the debate over the impact and value of strategic bombing has continued to draw battle lines within our own forces. Some proponents of air and space power claim that the advent of the latest capabilities of global attack and precision strike finally make independent strategic air warfare viable and desirable. As defined in JP 1-02, strategic air warfare is "Air combat and supporting operations designed to effect, through the systematic application of force to a selected series of vital targets, the progressive destruction and disintegration of the enemy's war-making capacity to a point where the enemy no longer retains the ability or the will to wage war." Those vital targets include "key manufacturing systems, sources of raw material, critical material, stockpiles, power systems, transportation systems, communication facilities, concentrated, uncommitted elements of enemy armed forces, key agricultural areas, and other such target systems."[2] Others argue that history, including the Gulf War, fails to demonstrate that strategic air warfare had the necessary effect on the will to wage war to achieve national objectives. However, the key merit in the newer capabilities for global attack and precision strike is their dramatic impact on the enemy's war-making capacity and ability.

At the operational level, the unique properties of global attack and precision strike allow U.S. air and space forces to shape the battlespace from a distance; to significantly influence the adversary's area of operation from outside his area and beyond his reach, thereby minimizing the placing of friendly forces in harm's way. The fewer U.S. forces placed at risk, the more credible are U.S. commitments and the more effective is operational employment. Our freedom of action creates operational opportunities while eliminating options available to the adversary.

Only air and space forces have the range, speed, deployability, and flexibility to engage rapidly, survivably, and accurately in any region. The unique contributions of air and space forces to global attack and precision strike are their capabilities to provide a quick and tailored response to areas essentially unreachable by surface forces. For example, when used in response to a cross-border invasion in a short-warning scenario, long-range air power equipped with smart munitions has the ability to halt enemy movement before critical territory is lost.

When used in a coherent, intense, joint forces campaign to make the adversary's armed forces and infrastructure militarily irrelevant, air and space forces could obviate high-risk, high-attrition surface battles, thus changing the composition and disposition of surface forces necessary to achieve military objectives. Shaping the campaign to U.S. design and constraining the adversary's choices allows U.S. forces to pursue full-spectrum dominance.

In the past, constraints on the use of force existed because of uncertainties about operational success and resulting effects. Precision strike capabilities increase the probability and predictability of success and decrease the prospects of casualties and collateral damage.

At the tactical level, precision strike decreases the forces and munitions necessary to destroy particular targets. Consider the trends in accuracy to achieve a 90 percent hit probability against a 60 x 100-foot target with a 2,000-pound bomb, as shown in table 4.1.

As shown in video from the Gulf War, precision strike now allows a single F-117 to deliver a guided 2,000-pound bomb into an area of an elevator shaft. Nonetheless, the true merit of precision strike is not just the increased efficiency in destroying targets. The impact of the increased efficiency is much more far-reaching. The benefits of fewer sorties and smaller munitions expenditures per target destroyed include reducing force packaging requirements, logistics needs, costs, forward basing, and war-fighters at risk. Or in the other direction, the benefits could

Table 4.1
Increased Efficiency through Precision Strike

War	Number of bombs	Number of aircraft	CEP* (in feet)
World War II	9,070	3,024	3,300
Korea	1,100	550	1,000
Vietnam	176	44	400

* circular error probable
Source: Richard P. Hallion, "Precision Guided Munitions and the New Era of Warfare," *Air Power History* 43, no. 3 (Fall 1996): 7.

be more targets destroyed. More dramatically, the implication is the capacity, especially with longer-range systems, to prosecute parallel attacks at the strategic, operational, and tactical levels simultaneously. Two years ago, the Air Force chief of staff, General Ronald Fogleman, put this in perspective:

> By comparison, during World War Two, the 8th Air Force attacked something like 50 target sets in all of 1943. During DESERT STORM, the coalition struck 150 individual targets in the first 24 hours. Not too far into the next century, we may be able to engage 1,500 targets within the first hour, if not the first minutes, of a conflict. Gone are the days of calculating aircraft-per-target kinds of ratios. Now, we think in terms of targets-per-aircraft.[3]

The revolution in C^4ISR promises a capability to match targets, weapons, desired accuracy, and platforms in near-real time in an iterative fashion. The conduct of a nonnuclear, largely precision campaign will employ near-real-time integration of advanced surveillance and battle management that connects and controls sensor to shooter architectures and facilitates the effective, efficient employment of joint air and space assets.

Challenges

Although the Working Group on Global Attack and Precision Strike is convinced of the unique merits of aerospace power as applied to the areas of global attack and precision strike, it is also concerned with the challenges that lie ahead in realizing those merits. Broadly, we believe that the competitive advantage promised by a global precision strike capability is threatened by inadequate or unbalanced investments. To maintain and expand that comparative advantage, we need to invest adequately in stealth, range, precision/near-precision munitions, C^{4}ISR, operator training and proficiency, mission planning, and the technologies enabling these capabilities. Specifically, we offer the following challenges as key issues to be addressed to ensure U.S. aerospace power's continuing capabilities through global attack and precision strike.

Among the extraordinary challenges posed by the capacity to execute a global attack and precision strike mission at the theater scale are the daunting C^{4}ISR and mission planning requirements. Consider the ISR man- and sensor-hours/days/weeks necessary to assess which are the most significant 1,500 targets for the outset of hostilities, what aim points would assure desired munitions effects, which targets require re-attack. Then there is the next round and the next days' targets, the C^4 capacity needed to plan, organize, safely execute, recover, and reconstitute participating air and space forces. These requirements would seem to belie the promise of rapid response for swift retaliation against an unexpected adversary.

The other major challenge is the education and adaptation of military and political policymakers to this new approach to warfare. The United States must alter the way it carries out its presence and power projection missions to meet the challenges of base denial/anti-access and asymmetric enemy responses. It is important to maintain the right balance between long-range bomber and other attack platforms. Only modern bombers can effectively attack deep into the adversary's territory from outside a theater of operations early in a conflict, or in the event that the bases and regional access required by other attack assets are denied. The investment in munitions needs to be better balanced against platforms, and better balance needs to be achieved between investments in direct attack and all-weather munitions compared with standoff and precision weapons. A greater emphasis on C^{4}ISR is needed to ensure system interoperability and to enable a global precision strike regime. These choices require an appreciation for the impact of air and space power on modern conflict. The battles for doctrinal preeminence, budgets, and specific programs can easily obscure the necessary focus on the best means to employ the modern instruments of military power. Even if the U.S. Air Force could "do it all," the question would remain, should it? The consensus of the working group is that the global attack and precision strike capabilities of air and space power establish the precursors to the successful employment of the remainder of the joint force team.

An ability to react rapidly with overwhelming effect to a crisis situation may not be easily transferred into action. The political leadership will need a clear understanding of the capabilities and inabilities of the military response. A decision to employ force in international affairs should not be hastened by the possessed capacity to respond rapidly. In some cases, time is of the essence, but the political fog, and the unintended consequences, may be much more difficult to ascertain than the military situation. Gone should be the days of the hair-trigger reaction to the use of military force. Nonetheless, the U.S. military should prepare the best means, and anticipate the resulting ends, to present the most effective options to the national leadership for its decisions regarding national objectives.

Opportunities

In discussing the merits of aerospace power and in developing challenges to the implementation of global attack/precision strike, the working group also noted the following issues and areas of opportunity.

Joint Issues

U.S. Air Force B-1s and B-52s can provide rapid power projection capability but are limited in their ability to penetrate without suffering unacceptable attrition. On-station carrier task forces could furnish fighter and electronic warfare (EW) support for the bomber force while USAF fighters are en route or slowed by unavailable access to a regional conflict.

How will future campaign planning leverage synergy between carrier aviation and the bomber force? Ground- or sea-based missiles offer a capability to complement air strikes. They can often respond faster, are less susceptible to adverse weather, and are less prone to losses from enemy air defenses. Greater emphasis needs to be placed on long-range nonnuclear ballistic missiles, including those launched from stealthy submarines or arsenal ships.

Force Modernization Issues

Modernization of global attack/precision strike assets is critical to maintaining a U.S. military advantage over any potential adversary. That said, these modernization programs must demonstrate a balanced investment in bomber and fighter aircraft, in standoff and direct attack munitions, and in the evolving C^{4}ISR architecture required to integrate these assets in a comprehensive precision strike regime.

U.S. Air Force Reserve and U.S. Air National Guard aircrews are at least as capable as their active force contemporaries, but reserve aircraft systems lag far behind in key munitions and EW capabilities. Thus, affordable modernization efforts should be extended beyond the active force.

Munitions Issues

The first issue in this category is the robustness of planned smart weapons delivery capability. USAF plans call for only five wings of aircraft with laser-guided munitions. All other aircraft will depend on near-precision weapons that may not perform well in a GPS-jamming environment. Low-cost terminal guidance, all-weather systems, synthetic aperture radar, millimeter wave radar, laser radar terminal guidance, and GPS counter-jamming techniques are approaches that need to be explored, compared, and pursued to provide a cost-effective and comprehensive mix of air-delivered munitions.

The second issue is smart weapons loading. Internal storage of weapons on stealthy fighter aircraft greatly limits their payload. Stealthy, conformal storage designs permitting low-observable outboard weapons carriage may make these platforms more cost-effective.

A third issue is bomb damage assessment (BDA). Off-board sensors, smarter munitions, "bell-ringers" built into warheads, or other technologies can improve BDA and make ground attack missions more effective.

Conclusion

Air and space power, enabled by the capabilities of global attack and precision strike, has come a long way over the last decade, and we expect progress to be even more dramatic in the new century. The merits of global attack and precision strike are that they present smaller, more survivable, and increasingly agile forces that have the punch to deter and defend but do not exhibit the vulnerabilities, infrastructure, expense, or risk of conventional Cold War air, land, and seagoing forces. These global attack and precision strike capabilities

- broaden the scope of options available to the National Command Authority and the war-fighting regional commanders;
- display an enhanced ability to project military power rapidly, with reduced risk, to any global trouble spot;
- guarantee American dominance in any future battlespace, whether or not the United States elects to commit substantial ground forces to a distant conflict;
- change traditional concepts regarding the mix of air and ground forces required to meet a major theater conflict;
- require new models of combat to be developed to accurately portray the contributions of air and space power; and, last,
- when combined in a precision strike regime, promise a dramatically new approach to underwriting future U.S. military commitments.

Notes

1. Department of Defense, JP 1-02, (Washington: Department of Defense, 1994), 302, 397, 411.
2. Ibid, 415.
3. Ronald R. Fogleman, "Getting the Air Force into the 21st Century" (remarks delivered to the U.S. Air Force Association's Air Warfare Symposium, Orlando, Fla., February 24, 1995), available at http://www.af.mil/cgibin/waisgate?WAISdocID=5929214620+2+0+0&WAISaction=retrieve; Internet; accessed June 7, 1997.

Note: The views presented in this chapter are those of the author as derived from the working group discussions and do not necessarily represent the views of the Department of the Air Force or the Department of Defense.

5

Information Superiority

C. Edward Peartree, C. Kenneth Allard, and
Carl O'Berry

Since ancient times, information has provided a decisive edge to warriors able to use it to their advantage. During the thirteenth century, Mongol "arrow riders" gathered and rapidly transmitted intelligence to field commanders across many miles, enabling them to exploit their superior mobility and cohesion to defeat much larger enemy forces. Similarly, superior U.S. information capabilities and the destruction of Iraqi intelligence and communications assets in the opening air campaign of Operation Desert Storm largely "decapitated" the Iraqi leadership, enabling the Coalition to exploit its superior mobility and cohesion to defeat the large Iraqi military force with minimal allied casualties.

Today, as the global effects of the Information Revolution expand and the U.S. military establishment contemplates the prospect of an information-based Revolution in Military Affairs (RMA), substantial levels of intellectual capital and other resources are being devoted to the question of how U.S. information superiority (IS) can be assured in future conflicts.[1] According to *Joint Vision 2010*, the vision of future conflict projected by the chairman of the Joint Chiefs of Staff, information superiority is the sine qua non of desired U.S. military capabilities. That blueprint postulates that any future military operation will be permeated with—and possibly decided by—information operations at all levels of conflict.

Revolutionary opportunities raise fundamental questions. Among the many issues raised by the RMA, this study considered the following:

- What kind of technologies and information operations are best suited to achieve the goal of information superiority that will provide the decisive combat edge to U.S. military forces at all levels of conflict?

- What changes in U.S. military organizations and institutions will be required to achieve the goal of information superiority?

- What are the implications for the nation's aerospace forces of both the goal of information superiority and the changes required to achieve it?

Beginning with an intellectual framework for future integrated information operations, the Working Group on Information Superiority examined the merits and roles of aerospace forces in achieving IS in terms of the information capabilities of all military services operating in a joint environment. The limits of aerospace-unique functions in achieving information superiority, acquisition trade-offs in acquiring information superiority, information system orientation (architecture, infrastructure, and interoperability considerations), technology vulnerabilities, and persistent issues concerning culture and institutions were also addressed. Finally, a set of recommendations was developed.

The Vision: An Integrated Concept for Achieving Information Superiority in Future Conflicts

With *Joint Vision 2010* as a conceptual capstone, a number of studies have examined the role of information operations and the importance of information superiority in future conflicts. It is this panel's contention that the Advanced Battlespace Information System (ABIS) study conducted by the Director, Defense Research and Engineering (DDR&E) and the Joint Staff in 1996 provides a viable roadmap for achieving IS in the future.[2] However, the integrated command, control, communications, computers, intelligence, surveillance, and reconnaissance (C^4ISR) infrastructure envisioned in ABIS and necessary to ensure information superiority at all levels of the joint environment does not yet exist. This vision of an integrated joint military information system is not an entirely new concept. Many of the same hopes were held out for earlier generations of communications technology: programs such as the Joint Tactical Communications Project (TRI-TAC) and systems such as the Worldwide Military Command and Control System (WWMCCS) and the Joint Tactical Information Distribution System (JTIDS) were incremental steps that ultimately fell short of their original goals.[3] The panel believes that although the ABIS vision and its associated technologies presume high levels of integration, certain other realities must be recognized:

- The tension between service and joint concerns.
- The incomplete revolution in applying commercial technology to military missions.
- The continuing profusion of systems, sensors and databases that often reflect the priorities of a single service or constituency.
- The need to come to terms with the austerities of future military budgets.

Largely because of these discontinuities, the undisputed information superiority now enjoyed by American combatant forces has been achieved primarily through the strength of the U.S. information technology industry, residual Cold War investments, and valiant efforts at the operator levels. The proliferation of divergent "stovepipe" systems has imposed burdens that often make technology seem as much enemy as friend to hard-pressed junior officers and non-commissioned officers (NCOs). Consequently, there is a real need to make technology choices that not only ease this burden but also reduce the discrepancies between different operating systems—and between those systems and the familiar products of the commercial marketplace.

Achieving the technological vision of *Joint Vision 2010* and ABIS in the face of these continuing realities will require a significantly different way of thinking and doing business: streamlined processes for integrating joint doctrine with operational experience, redefined roles and missions, and an unyielding commitment to focused technology investment and joint operations. Only then will the ideal of a globally available "system of systems" in support of U.S. military forces be realized.

New Operational Concepts Enabled by Information Superiority

Although the utility of information in war is not new, the speed, scope, and overriding importance of information is at the heart of the RMA. Information superiority is basic to the four pillars of *Joint Vision 2010:* dominant maneuver, precision engagement, focused logistics, and full dimension protection.

Each of these pillars contributes to a decisive U.S. advantage in future conflicts through "full spectrum dominance." *Dominant maneuver* leverages information to synchronize forces for rapid mobility and control of the battlespace. *Precision engagement* demands accurate, timely, and prioritized target acquisition, as well as command and control features for maximum effectiveness in engagements. *Focused logistics* fuses information, transportation, and supply requirements to deliver tailored logistics packages to fielded forces. *Full dimension protection* requires tailored, timely, actionable information on threats to ensure protection of all assets within the battlespace.

What is clearly new about this vision is *the extent to which it relies on information as a weapon of war* that U.S. forces must wield effectively in order to prevail. It follows that, more than any other weapon in the American arsenal, the information weapon is inherently joint, even though it depends on discrete contributions from each of the services and many national agencies. This shared awareness is fundamental to the integrated joint teamwork needed to achieve decisive results at any point on the spectrum of conflict. Intelligence focused on the warrior or the peacekeeper will mean fewer bullets expended, lives lost, and resources consumed. Whatever the mission, IS implies a faster, more precise ability to terminate engagements on terms favorable to U.S. interests. No single service is by itself responsible for or capable of achieving IS: The full spectrum dominance required by *Joint Vision 2010* therefore demands an integrated joint effort.

Achieving information superiority will have wide-ranging effects on the way that American forces plan and conduct military operations. These highly interdependent effects include the following:

- With common battlespace awareness and assured real-time information, the tempo of combat can be greatly accelerated.

- With a streamlined, highly networked information dissemination and retrieval system, war-fighters at the tactical level can be empowered with greater freedom of action because they have timely, relevant information on the battlefield.

- With real-time sensor to shooter coupling enabled by wide band links and intelligent data bases, fewer critical nodes are manned by humans.

- Concentration of fires through the increased precision of information-enriched weapons makes the battlespace more lethal and, necessarily, more dispersed. As a result, military forces and their support structures (including command and logistics) must be far less concentrated than ever before.

- This also implies a reduction in numbers of platforms (many of which will be unmanned) and a concomitant increase in the information content of remaining platforms. Precise long-range fires can be delivered from remote air-, land-, or sea-based platforms.

- Because of these qualitative changes, the eternal balance between centralized command and decentralized execution will be more important—and even more difficult to achieve.

Rigidly centralized control (micromanagement) and total decentralization (anarchy) are equally self-defeating, and technology by itself will not be able to maintain this balance. As in any other investment choice, risks and their implications must be managed through a combination of focused leadership and coherent strategic choices. Those choices must begin with an understanding that achieving information superiority depends on the ability to leverage the national information infrastructure—of which the military is only a part.

Information warfare (IW) is therefore a critical element in this equation. Defined as "information operations conducted during time of crisis or conflict to achieve or promote specific objectives over a specific adversary or adversaries,"[4] IW has both offensive and defensive applications within tactical, operational, and strategic dimensions. *Strategic IW* includes the targeting (offensive) or protection (defensive) of homeland assets such as information and internetted critical national infrastructures. *Operational IW* includes the ability to target or protect information assets at the level of campaigns. At the tactical level, IW entails the ability to disrupt, disable, deny, or exploit enemy battlespace information capabilities and protect one's own.

Protecting our own critical systems is essential to maintaining information superiority. Protecting systems and assuring information requires strong measures for defensive IW: strong encryption, intelligent firewalls, and other countermeasures, integrated with such traditional skills as personnel and physical security. Offensive IW must include kinetic attack capabilities directed at critical infrastructure nodes (as shown during Desert

Storm) as well as digital attack capabilities to enhance the traditional electronic warfare mission (jamming and exploiting adversary signals).

Finally, information superiority, when realized, promises to enhance the effectiveness of the joint force. Seamless interoperability of heterogeneous systems across and within services is a precondition to the achievement of IS. A common, consistent awareness of the battlespace would promote the synchronization of forces and maximize the use of military assets across services to accomplish goals and exercise decisive advantage. The overriding goal here is to end the spatial boundaries and "turf" that formerly marked off the battlespace among the services.

As an element of statecraft, information superiority may also facilitate conflict avoidance by bringing advance knowledge of adversary capabilities and intentions. This concept is enabled by a number of emerging technologies including advanced synthetic aperture radar deployed on Joint Surveillance and Target Attack Radar Systems (JSTARS), hyperspectral sensors on unmanned aerial vehicles (UAVs), and adaptive optics satellite technologies with ever-improving resolution, all supported by highly fused knowledge-bases and worldwide data dissemination systems. Any mobilization, or attempt at covert strike, or violation of arms control or treaty arrangement could be detected and, ideally, deterred. The speed of communications links ensures that the relevant information can reach those who need it, that warnings can be issued quickly, and that timely actions can be taken. The speed of information gathering and flows may thus allow the management of crises and the mitigation of risks. This is an intriguing vision for the future, but, as always, the most significant limitations in achieving it are human factors—individual, organizational, institutional, cultural, and political.

The Aerospace Vector

Of all the services, only the U.S. Air Force (USAF) has elevated the concept of information superiority to such a level of importance that it can be called a core competency.[5] Indeed, since the first use of military balloons to elevate the "high ground" and thus improve awareness during the French Revolution, aerospace forces have been key assets in accomplishing the intelligence/surveillance/reconnaissance mission. With new sensor, fusion, and dissemination technologies, and with the advent of airborne command and control, aerospace forces have expanded their competencies to cover the entire C^{4}ISR mission. The USAF's Rivet Joint electronic signals monitoring, airborne warning and control systems (AWACS), and JSTARS have proven themselves essential platforms in conflict areas supporting information requirements. New generations of information-gathering and disseminating UAVs such as Predator, Global Hawk, and Dark Star are being developed and given oversight by the USAF through the Defense Airborne Reconnaissance Office (DARO). The preponderance of spaceborne defense reconnaissance, communications, and geolocation capabilities are under the executive agency of the air force. Each of these air- and space-based assets is a fundamental hardware component of—and provides content products for—an emerging "system of systems" that will facilitate twenty-first century information superiority: the global command and control system (GCCS) coupled with the global broadcast system (GBS). The ability of other services to do long-range remote fires (assessed in the Deep Attack Weapons Mix Study [DAWMS]) rests on the availability of USAF information-gathering and dissemination capabilities.

The U.S. Air Force not only has had a historic role in developing intelligence-gathering, communications, and command and control systems, but it has also been at the leading edge with regard to the development of doctrine and operational concepts involving information warfare and the achievement of information superiority. The Air Force Information Warfare Center (AFIWC) located at Kelly Air Force Base, Texas, was the first such center to explore the issues and game the problems of IW in the new and future conflict environment. The USAF further demonstrated its commitment to the primacy of information superiority by creating the 609th Information Warfare Squadron at Shaw Air Force Base, South Carolina. Further evidence of the commitment to competency in IS is seen in the prominent place given information superiority and information warfare in the USAF's near-term vision document *Global Engagement* and the longer-term vision document *New World Vistas*. Aligning structure with its future IS vision, the U.S. Air Force has been the first military service to integrate the intelligence and operations functions within its headquarters staff.

Precisely because of this institutional commitment and its technological core competencies, the panel feels strongly that the air force is uniquely positioned to play a vital role both in capturing the new vision of information superiority and in solving the discontinuities outlined above. Far from adhering to traditional practices of using these technologies to fulfill service-specific roles and missions with joint interoperability as a secondary concern, the air force can build on its core competencies to help provide a major part of the institutional architecture of the future with joint IS as the principal goal. This vision of the service as the central provider and facilitator of a new family of joint IS capabilities frames the discussion of specific issues that follows.

Key Issues

A diverse set of problems is associated with achieving information superiority and effectively implementing the vision that has been laid out above. Technological/infrastructural, architectural, institutional, cultural, budgetary, and human limitations impede progress from vision to reality. Of paramount importance is a robust national information infrastructure (NII) and global information infrastructure (GII). These are necessary to ensure the viability of tactical, operational, and strategic information capabilities required by military forces seeking information superiority. It is generally accepted that as much as 95 percent of military communications travel over commercial systems. Therefore, there should be keen air force interest in seeing that the emerging concept of *convergence*—which refers to the merging of global communications infrastructure, information transfer processes (wire, fiber-optic, wireless), and computer operations and services into a vast, highly robust digital information environment for satisfying all user needs virtually instantly—comes to fruition.

Historically, early adopters and integrators of technological innovations have been frequently rewarded in both business and war; those who have been slow to adapt and integrate have often suffered grim consequences. This warning underscores the seriousness of these issues; given the present pace of technology advances in the commercial market, time becomes the critical factor in matching military information system investments with organizational decisions. Consequently, we emphasize that the military must turn immediately and unhesitatingly to the commercial sector to upgrade its information systems and capabilities.

Technology/Infrastructure

The overarching technological/infrastructural issue is the need to ascertain and achieve the correct balance between commercial and purely military systems and applications. The technology of military information systems must match the leading edge in the commercial marketplace, while allowing limited military-unique capabilities. For example, whereas commercially available information pipes are sufficient to provide information inter-theater, operation in a hostile intra-theater environment requires broader bandwidth than is likely to be available in most near-term commercial systems. Also, in many instances, it is essential to "ruggedize" and harden information systems and computer hardware in order to offset combat conditions. Simple environmental factors like Bosnian mud and desert dust have played havoc with computers in recent deployments and war games and sorely reduced their utility.[6]

The difficulty of information sharing and system security, particularly in coalition operations, is a persistent problem requiring serious attention in the system acquisition process. System robustness and survivability also continue to be of concern, although a truly global infrastructure is beginning to emerge, offering the promise of immense redundancy and virtual connectivity that, coupled with new security certification and authentication techniques, should simplify information sharing among frequently changing coalitions.

Ground stations in theater will continue to be vulnerable to various kinds of attack until adaptive networks that eliminate or minimize vulnerable nodes become a reality. Satellites on which geolocation, data collection, and information dissemination critically depend are potentially vulnerable to electromagnetic pulse (EMP) and other antisatellite weapons; hardening them against attack is expensive and defensive weapons are currently not a Department of Defense (DOD) priority. Data fusion—the capability to effectively integrate information from multiple sources to create a complete picture of the battlespace—is not yet sufficiently robust. And the management of large, heterogeneous systems, though not an insurmountable undertaking, is still very difficult.

Architecture

What might once have been technology issues—bandwidth, computing speed, memory capacity, data storage and dissemination—have largely been supplanted by architectural issues. Although the technology is now available, the Department of Defense, driven by concern for retention of the functions resident in legacy systems, currently lacks the required architectural vision to integrate new technology into a dependable "system of systems" capable of satisfying the requirements for information superiority in a joint battlespace. Future integrated battlespace information concepts emphasize that information structures and flows must be addressed as a continuum all the way from the individual war-fighter (tactical operations) to the National Command Authority (strategic and global domain). Recent evidence from Bosnia suggests, however, that "the information revolution stops at the division level."[7] An interoperable, highly redundant, and robust system oriented to individual user needs and based on commercial standards is essential.

Architecture versus Platforms

Historically, too much emphasis has been put on the performance of stand-alone platforms and too little on the value of adding the platform to an overall system. Focusing on the overall system—the architecture—as opposed to the unique capabilities of a given platform is essential to achieving information superiority. The architecture for military information systems must come from the chairman of the Joint Chiefs of Staff; however, absent fundamental change in the budgeting, planning and programming processes, the implementation can only be done by the Armed Services.

Institutional Concerns

Institutional issues may be the thorniest of all to resolve. The acquisition process and the organizational culture of parochialism prevent the services and DOD from pursuing truly cooperative movement toward integrated information systems that can support joint forces and joint operations.

The traditional acquisition process is optimized for the procurement of large platforms with military-unique functions over the course of multiyear cycles. Such an acquisition process has proven increasingly ineffective in adapting to an information-based RMA driven by commercial technology, with multiple technology generations operating within a single DOD acquisition cycle. Military operations in Somalia, Bosnia, and

Chechnya have amply demonstrated that irregular forces, employing off-the-shelf information technology—particularly cellular communications—have enjoyed better communications than the highly industrialized military forces they faced. The current acquisition process, despite substantial reforms in law and regulation, leads to modeling, simulation, computing, and telecommunications products that are obsolete before they reach the field and to service-oriented architectural stovepipes—serious obstacles to achievement of unequivocal information superiority.

Thus the organizational culture of the services and DOD is an impediment to the pursuit of information superiority. Parochial interests preserve an acquisition system oriented toward large platforms and traditional military hardware. New-generation UAVs and information systems lack critical constituencies to push them through those processes. A deeply embedded culture of risk avoidance impedes experimentation with, and employment of, faster acquisition and integration cycles, making commercial off-the-shelf (COTS) procurement of information systems far more difficult.

Changing deeply embedded cultures requires a shift in focus from a service orientation to jointly integrated missions. Much of the implementation of the joint vision is left to the services, each of which must compete with the others for roles, missions, and dollars. Given the widely acknowledged significance of information superiority in the battlespace of the future and the USAF assertion of IS as a core competency, a different focus on information technologies (IT) and information operations is essential.

Budgetary Concerns/Acquisition

In an environment of shrinking defense budgets, fiscal issues tend to center on difficult trade-offs between firepower, manpower, and information systems. One of the key problems in promoting information system purchases is that it is hard to measure how information systems contribute to the fight (as compared to conventional measurement of platform/firepower asset performance). These inherent difficulties reflect the fact that information systems tend to be catalytic and integrating tools—offering the prospect of improved efficiency and effectiveness—as opposed to stand-alone weapons that can be quantitatively tested. Advanced concept technology demonstrations (ACTDs) offer the possibility of demonstrating the advantage of investment in information superiority, but the results of such demonstrations are not widely applied. Worse yet, they are now being substantially reduced.

Human Concerns

The limits of human cognition and the risks of information overload pose significant challenges where a premium is put on decentralized decision-making, self-synchronization, and adaptation to fast-moving conditions in an information-rich environment. There is a real, persistent danger in creating a situation where the proliferation of irrelevant data can overwhelm the war-fighter. Prioritization of data collection, fusion, and dissemination issues must be worked out to ensure that relevant information reaches the right person in a timely fashion.

The balance to be struck between autonomy and control is critical. The success of new operational concepts envisioned in an environment in which U.S. forces enjoy information superiority depends on a decentralized command environment where individual war-fighters are empowered to make decisions supported by access to all relevant information. Whereas command and control functions will remain hierarchical, information flows must be networked. Much greater emphasis must therefore be put on the principles of "commander's intent" and "control by negation" (e.g., a war-fighter seizes the initiative unless specifically directed *not* to do so).

The problem is less one of technology than of focus and integration; instead of concentrating on new systems, infrastructure, and bandwidth, emphasis must be placed on the individual war-fighter and his or her needs. Reducing discontinuities between military and commercial standards and creating an architecture that integrates systems are essential. Not only would the resultant systems be more responsive, but also the overall architecture would significantly reduce vulnerable nodes through redundancy and natural robustness.

Conclusions and Recommendations

Information superiority is the cognitive high ground in future conflict. Not only does it enhance military capabilities across the spectrum of conflict, it is essential to the functioning of air and space power itself. All five requirements for effective air power (navigation, targeting, intelligence, effects, and survivability) can be attained only when information superiority is achieved.

IS will also become essential to the growing dimension of space power; currently space-based capabilities are a key provider of the components of information superiority.

Most important, information superiority, enabled by information technology, will permit effective integration and optimization of all U.S. military forces. It is apparent that the U.S. Air Force specifically, and DOD generally, are initially on the right path or "vector" to achieve information superiority. The United States sets the world's standards for the global information industrial base in terms of hardware and software production, and the military services draw from the most computer-literate society in the world. Achieving the full potential of information superiority, however, is dependent on sustaining consistent effort along the present USAF vector and helping this thinking to migrate across all of the Department of Defense.

While the vector to achieve information superiority is correct, there are difficult choices/trade-offs ahead. One of the first issues to be addressed will be developing a "user-oriented" architecture based on military systems aligned with superior commercial systems as the global grid approaches convergence. Additionally, there must be balance between information offense (attack), defense (protect), and exploitation (intelligence gathering). In the past exploitation has drawn the majority of the available resources, and current resource allocation trends favor offensive operations. System vulnerabilities and the strong requirements for information assurance (protection of the privacy and quality of transmissions) will have to be addressed and resolved if information superiority is to be achieved and sustained. Further, sustained commitment to investment in new information-based programs, battle laboratories, and robust joint exercises will be required to make the IS concept into a reality. Strategic vision, sustained leadership, resource allocation, and a commitment to jointness will be essential.

The Working Group on Information Superiority believes that the U.S. Air Force clearly has commitment from its top leadership, is focused on the most critical areas, and is on the correct vector to achieve its goals in information superiority and information operations (IO). The panel believes that the air force could accelerate its progress with IS by taking the following actions:

1. Coordinate information operations across services and national intelligence organizations to achieve informa-

tion superiority and use operational Red Teaming/Joint exercises to realistically test technology and capabilities.

2. Adopt, wherever possible, commercial acquisition processes for information technologies using a risk management rather than a risk avoidance philosophy. This includes experimenting with different systems and innovating based on commercial practices. Adjust the acquisition process to facilitate the purchase of information technologies based on the evolving global commercial architecture, and embrace "convergence" as the principal guiding concept for architecture and infrastructure development.

3. Commit to an "operator-oriented/user-friendly" approach to building infrastructure and architecture to help reduce discontinuities between military and commercial systems.

4. Align the USAF with joint vision. Information superiority is for all services at all echelons. It is also recommended that all services follow the USAF lead in committing to information superiority as a required core competency.

5. Establish and document in regulatory guidance processes to define, validate and evolve IS/IO requirements within the USAF.

6. Use USAF processes to input validated operational requirements into the Joint Requirements Validation process.

7. Define and establish metrics to measure/document progress.

8. Commit the entire USAF to a process to "educate, test, and grade" IS/IO. Recognizing that these are cultural changes and that experience is the best teacher, the tests (exercises) for information superiority operations must use real-world sensors/assets at the tactical and national levels to establish the desired experience level. Additionally, advancement/promotion must be contingent on how well one performs in this new environment at the personal, group, and command levels. This recommendation is analogous to the Goldwater-Nichols

legislation that made senior officer promotion contingent on successful joint duty. This legislation caused the best officers in the services to seek joint assignments.

The present commitment from the top leadership in the U.S. Air Force must be sustained over time due to the magnitude of cultural change required to make information superiority a core competency. In the Armed Services, the service chief is recognized as the head strategist, operator and resource allocator in uniform. The U.S. Air Force has moved dramatically to achieve cultural change and greater capabilities in information superiority by redefining the term "operator" to include "information operations." Unless there is a sustained commitment from the senior leadership of the USAF, IS/IO will devolve back to a second- or third-tier supporting service as opposed to a core operational competency.

Notes

1. *Information superiority* is defined by the Department of Defense as "the capability to collect, process, and disseminate an uninterrupted flow of information while exploiting or denying an adversary's ability to do the same." Department of Defense, DODD 3600.1, *Information Operations*, December 9, 1996.

2. Director, Defense Research and Engineering and Joint Staff C^4 Directorate, *Advanced Battlespace Information System*, Task Force Report, vols. 1–6 (Washington, D.C.: Department of Defense, 1996).

3. C. Kenneth Allard, *Command, Control, and the Common Defense*, revised ed. (Washington, D.C.: National Defense University Press, 1996).

4. Department of Defense, DODD 3600.1, *Information Operations*.

5. The USAF lists information superiority as one of its six core competencies (the others being air and space superiority, global attack, global mobility, precision engagement, and agile combat support. Department of the Air Force, *Global Engagement: A Vision of the 21st Century Air Force* (Washington, D.C.: GPO, 1996).

6. Besides problems with dust, recent accounts of the U.S. Army's digital Advanced Warfighting Experiment (AWE) held at the National Training Center have also indicated difficulties with excessive information causing systems to crash and an environment so full of electronic communications that radios ceased to function. Mark Thompson, "Wired for War," *Time*, March 31, 1997, 72.

7. Kenneth Allard, "Information Operations in Bosnia: A Preliminary Assessment," *INSS Strategic Forum*, no. 91 (November 1996).

6

Mobility and Support

Keith A. Hutcheson and Robert McClure

> *"If we do not build a transportation system that can meet our needs of tomorrow, then it doesn't matter much what kind of force we have because it won't be able to get there."*
> —Gen. John M. Shalikashvili

In a world that has rapidly evolved from bipolar to that which exists today, our national security strategy of "engagement and enlargement" and the national military strategy of "prevent, deter, defeat," have taken on new significance and clearly mean that America will remain involved in the world. Our forces are increasingly based in the continental United States (CONUS) and have been dramatically downsized. Yet we conduct humanitarian and peacekeeping operations, in locations rarely visited before. The speed of reaction has become almost as important as the desired outcome. America must have an enabler to help maintain its influence and power across the entire spectrum—from prevention to deterrence to defeat of an adversary. Mobility, and most important, *air mobility*, is that great enabler. An integrated system of platforms, people, information, and support, air mobility is often not recognized as an essential element to American foreign policy and our national security strategy.

Air mobility has been overlooked for more than 50 years despite the magnitude of its accomplishments. The Berlin Airlift deterred a probable conflict without firing a shot. More recently, U.S. air mobility rushed forces to the Persian Gulf in 1994 to deter a second Gulf War. Almost daily one reads of air mobility assets assisting humanitarian operations throughout the world. Clearly, without air mobility America would be hard pressed to be the dynamic world power it is today. Our resolve, recognized by friend and foe alike, to commit the nation to action is best illustrated by the daily activity of airlift and tanker forces worldwide in operations ranging from humanitarian relief to delivering deterrent forces to a threatened region. Quite simply, air mobility is the guarantor of the American ability to apply force whenever the nation wishes, anywhere in the world.

The Case for Air Mobility

The CNN image of Congressional Medal of Honor winner Staff Sergeant Randall Shugart's body being carried off the back of a C-141 at Dover Air Force Base was not only gripping, but also perhaps one of the few glimpses the American public ever gets of its air mobility fleet. Built to rapidly deploy and sustain combat forces overseas, the wings of C-5, C-17, C-130, C-141, KC-10, and KC-135 aircraft have evolved to become essential elements of our national military strategy as well as the means by which heroes are returned home from far-off places like Mogadishu, Somalia. This image will become even more ironic when the *USS Shugart*, America's latest heavy-force deployment ship, built to rapidly move decisive force around the world, is commissioned in 1997 in his honor. Indeed, from our military history for just the past 50 years, a significant case can be made that the speed and timing with which our nation has rushed force to places of vital interest has had an inverse effect on the number of coffins returned to the United States. Following North Korea's attack across the 38th parallel in 1950, it took weeks to move the first combat forces from America into that war; the memorial to more than 54,000 who died there was unveiled in Washington, D.C., only last year. More recently, air mobility moved a Patriot unit from Germany to Israel during Operation Desert Storm in less than 22 hours, thus keeping that conflict from spreading with dangerous consequences. Fewer than 50 American combat casualties returned to Dover from Desert Storm, as SSG Shugart would later from Somalia, in part because of the increased importance and role all, but particularly air, mobility forces played.

Think for a moment about the United States without its air mobility capabilities. It would be like most other nations. We would not be able to use commercial passenger and cargo aircraft to move troops and supplies into permissive environments. We would have to rely on locally available vehicles and major equipment until a ship arrived because no outsize cargo could be flown. Our ships could sail to many parts of the world but would have little staying power. Without in-flight refueling our fighters could deploy great distances only by island-hopping or on carrier decks. In essence, our international status would depend on economic and diplomatic influence without the ability to back up commitments in times of conflict or to relieve human suffering. Clearly, this picture is not one of a great nation.

Gen. Ronald Fogleman, former U.S. Air Force chief of staff, called strategic air mobility "the linchpin of National Military Strategy."[1] Indeed, because of its ability to project forces, our air mobility system holds the various elements of U.S. military strategy together. Former Secretary of Defense William Perry observed that this strategy was built around three words: prevent, deter, defeat. By its daily demonstration of American military and political power around the world, air mobility provides a level of energy to our nation's resolve and commitment; without it, we would be marginalized across the entire spectrum from prevention of conflict to defeat of an enemy. A rapid global transportation system is absolutely necessary, even critical, to quickly back up our national security and military strategies. The air mobility system provides the timeliness and capacity to meet our national needs. The elements of this system include not only airlift, but also air refueling assets and the global support structure of people and equipment that allows the system to function efficiently and effectively.

The certain uncertainty of the strategic environment is problematic. The United States must respond to this unique and challenging environment to remain a dominant force in the evolving new world order.

On the one hand, the dissolution of the Soviet Union has been destabilizing; militant Third World nations are emerging; and a global proliferation of technology, including WMD, has increased the lethality of potential conflicts. Coupled with this new world order is the severely constrained U.S. defense budget that is directing the downsizing of our forces and consolidation in CONUS. On the other hand, today, unlike in the Cold War era, there is a greater probability of conflict as the fear of starting World War III has diminished and many smaller states and non-state actors have begun to flex their military muscles. Consequently, an increased number of small conflicts, as well as peacekeeping and peacemaking efforts, are typical of today's environment, but these are also highly unpredictable.

Grenada, Panama, Haiti, Somalia, and Bosnia have all been neither typical nor predictable. Conversely, the confrontation between NATO and the former USSR as well as a second Korean War did not materialize. Moreover, today's conflicts generally appear on short notice. It is difficult to tailor forces and respond to these unforeseen contingencies. A smaller overseas presence and a reduced, or perhaps misplaced, pre-positioning of stocks only aggravate this situation.

The air mobility system's inherent flexibility and adaptiveness is a good news story. Although the problems are daunting, they are not insurmountable. As this essay will illustrate, the mobility forces of the United States, in particular *air mobility*, provide the timely leverage needed by our political leaders to confront these challenges and to thwart aggression and promote stability.

The Elements of Air Mobility

Planning

Historically, war planning assumptions have been very optimistic about the availability of air mobility, yet pessimistic in regard to other weapons systems. In actuality, we tend to have an overcapacity to kill targets yet a limited ability to move these killer forces to engage the enemy. Over the past 15 years, projects such as the 1981 congressionally mandated mobility study, the 1995 Mobility Requirements Task Force update of the Bottom-Up Review (MRS BURU), the 1996 Defense Science Review Board Mobility Summer Study, and the 1997 Intra-theater Lift Analysis have determined that air mobility forces may *not* be available when needed. The MRS BURU states that, to fight two nearly simultaneous major regional conflicts, the demands placed on air mobility would exceed the lift available. Additionally, war planning assumptions may be well short if we fail to account for the day-to-day peacetime requirements that send air mobility forces throughout the world. Reassembling these dispersed air mobility forces takes precious time. Finally, war planning assumptions do not always address the political dilemma of turning away from ongoing humanitarian and peace operations (Bosnia, Southwest Asia, Rwanda, etc.) to respond to regional warfare.

The System

Air mobility, which includes airlift and aerial refueling, is the foundation of the mobility triad. All three elements of mobility provide unique qualities. Pre-positioned (PREPO) stocks allow for forward-basing of equipment, and sealift provides the bulk of follow-on equipment. Air mobility is fast and agile. The merits of air mobility are threefold: it is a national asset, it provides political leverage, and it is a force enabler. Air mobility is the only leg of the triad that brings the right people and the right equipment to the fighting force in a timely fashion.

The potential rapid employment of U.S. military forces—"anything, anywhere, at anytime"—is the fist held in reserve to uphold and defend American interests around the world. States intent on threatening these interests are less likely to do so if they know we can respond quickly and decisively. As the new world order has emerged, uncertainty and lethality implications appear to dictate that the ability to move robust combat forces rapidly will become essential to prevent crises from spinning out of control. Air mobility is the means to rapidly deploy and employ these forces—it is the *backbone of deterrence*. It is the mechanism for rapidly moving people and equipment from theater to theater and within theaters.

Air mobility is not theater-specific; it is everywhere, in all environments and weather conditions, in peacetime and in war. Since the acquisition of the C-17, the traditional airlift categories of inter- and intra-theater have become blurred with direct delivery to forward locations. Assets can now perform inter-theater and intra-theater functions. Fewer "seams" will further expand the flexibility and adaptiveness of future air mobility forces.

Aerial refueling assets are an essential element of the air mobility force. Their most obvious attribute is extending the range of American and allied aircraft, thereby increasing the agility and flexibility of airlift aircraft. In the employment segment of a scenario, aerial refueling forces become force enhancement tools highly sought by the theater commander. In addition, aerial refueling tankers have the capability of carrying cargo and personnel whether they are refueling or not.

In Total

Air mobility focuses the global transportation system. Air mobility is much more than the primary facilitator (aircraft); it is a complete closed-loop system that includes command and control, planning and execution, and the needs of the theater commander in chief. If air mobility shuts down, the entire fighting force loses focus. Therefore, we must aggressively pursue the modernization of air mobility forces to meet the readiness support needs of training, personnel, equipment and spares. As long as air mobility is planned efficiently and not overly optimistically, it will continue to serve our nation and the world well.

Where Does Air Mobility Fit?

Air mobility is the primary means for rapidly deploying and sustaining forces. The global transportation system is often misunderstood because of its complexity and dynamic nature. It is, however, the key element of the overall military logistics system. *Logistics* is best described as a system comprised of complementary and interdependent functional parts. The mobility triad is a facet of the transportation variable in the logistical equation.

Metaphorically, mobility is like the end of a dog's tail. The dog's shaking and moving is exaggerated at the tip of its tail. Current and long-range initiatives within the logistical system—the shaking and moving—particularly within "support," will have a direct impact on mobility operations—the end of the tail. Support initiatives such as lean logistics, assured delivery, velocity management, and time-definite delivery are all efforts to improve the overall effectiveness of the global logistics system by reducing cycle times and supply stocks. But, our air mobility force is agile enough to accommodate changes as they develop and flexible enough to alter traditional methods of operation to meet new challenges.

Underlying Assumptions

How does air mobility make the overall military transportation system more effective? Effectiveness in the rapid global mobility world translates to faster closure times for forces. Before we can articulate the merits, however, we must make certain underlying assumptions about available resources. The following resources, though not all-inclusive, must be recognized, provided, and maintained if we hope to gain the potential benefits of the air mobility system:

- Sufficient personnel, operations, and support, to meet mission needs. The time people are deployed (PERSTEMPO) must be kept at a realistic rate. Indeed, PERSTEMPO, or personnel tempo as it is called in the military, is the "long pole in the tent." If people are too tired or their morale is too low, we cannot expect them to perform their task properly which jeopardizes the overall mission.

- Adequate numbers of qualified aerial-refueling and aerial-delivery crews.

- Sufficient en-route infrastructure, support, personnel, and equipment at staging, en-route, and off-load location stations. Some flexibility (i.e., swing capability) will be lost as fewer C-17s (120) are procured to replace large numbers of retiring C-141s (248).

- Greater interaction and communication between U.S. Transportation Command (USTRANSCOM), commanders-in-chief (CINCs), and theater transportation organizations in deliberate planning and crisis execution. This includes a more disciplined use of the transportation system in general.

- Continued modernization of global and tactical information support systems.

- Continuation of initiatives to protect sea/aerial ports of debarkation (S/APOD) and air mobility assets from weapons of mass destruction and advanced conventional weapons.

- Readiness accounts for training and support equipment (spares, forklifts, K-loaders, etc.) that keep pace with needs.

- Minimized handling and transfer of personnel and equipment.

Air Mobility Meets the Challenge

Anthony Cordesman, in "The Quadrennial Defense Review and the American Threat to the United States" stated, "it is doubtful that any United States strategy can prevent the United States from having to react to new and often unanticipated military challenges."[2] We, indeed, are a reactive force. As such, it is difficult, if not impossible, to tailor an appropriate force required for a given scenario. So the questions is, how do we bring the proper force to theater in the quickest time? The answer is Global Air Mobility.

The United States and the rest of the world depend on air mobility far more than most people realize. If not for air mobility, British bombers would not have obtained aerial refueling support on their way to the Falkland Islands. If not for air mobility, literally, thousands of people in Rwanda, northern Iraq, and Somalia would have died of starvation or deprivation. If not for air mobility, forest fires in the southwestern United States would have destroyed far more than they did. Air mobility delivered the halting mechanism that prevented Saddam Hussein from invading Saudi Arabia. Air mobility is everywhere, everyday, defending the interests of the United States and the international community.

Air mobility has numerous merits, but the most significant ones can be divided into three basic categories. First, air mobility is a *national asset*:

- It is *responsive*. Increased uncertainty and instability around the world demand a responsive and air-delivered force. America's interests abroad can be defended and maintained only with a rapid response capability. Air mobility fits that bill, particularly as there are fewer and fewer forward-deployed forces everyday.

- It is a *balanced force*. The air mobility system incorporate active duty, air force reserve (AFRES), and air national guard (ANG) forces as well as private industry into its mix, offering a wide range of experience, breadth, and talent. The Civil Reserve Air Fleet is clear evidence of that cooperation.

- It is *priority allocated*. Strategic air mobility assets are national assets. Priority use is allocated by the Joint Chiefs of Staff based on the mission assigned by the National Command Authority.

Second, air mobility is a *political asset:*

- Air mobility is a way to display American resolve and commitment to its interests and allies. It also serves as a deterrent to potential adversaries through its demonstrated presence, while helping to build goodwill worldwide. The speed and range of air mobility allows our political leaders to respond decisively to crises anywhere. When humanitarian emergencies arise, air mobility is often the first to respond. (e.g., in Bosnia, Caribbean, and

Zaire). The ability of the C-17 and C-130 to operate in and out of austere and expeditionary airfields widens the reach of military forces and humanitarian aid.

Third, air mobility is a *force enabler*. Its features in this category include

- *Unimpeded movement.* Air mobility projects power by transporting forces and materiél rapidly without regard to surface obstacles.

- *Force multiplier.* The aerial refueling capability of air mobility ensures that the unique flexibility of air power concentration—at any point on the globe, at a point of our choosing—will remain a reality. Aerial refueling expands the operational realm of U.S. and allied aircraft by extending their payload, range, and endurance. These assets act as a force multiplier by extending U.S. reach as well as that of our receiver-capable allies, particularly now that fewer overseas bases have refueling facilities.

- *Combat insertion.* Air mobility provides the capability to deliver airborne combat forces right in the face of the enemy.

- *A new way to the fight.* The C-17, with its outsized cargo capacity and its extended range over the C-130, brings a new dimension of agility to air mobility that minimizes the problem of seams in the transportation system. Agile air mobility assets can deploy, employ, and sustain dispersed forces accurately and timely while reducing the amount of logistics and the number of personnel required in theater.

It must be remembered that air mobility, although it is the foundation of the global transportation system, complements the synergistic and interdependent pre-positioning of stockpiles and the movement of forces and equipment over water.

Conclusion: Future Demands on Air Mobility Will Only Increase

Our air mobility forces must continue to modernize their airlift and tanker forces, upgrade global traffic management to Federal Aviation Administration standards, enhance their support structure and equipment, and provide the information systems for real-time command, control, and in-transit visibility. The demands placed on our global transportation system will continue to increase in the foreseeable future. Why? As denizens of the global village we have come to expect—indeed, to demand—a quick and decisive response to crises around the world. Advanced communications technology, which has "shrunk" the world, brings the images of these crises into our living rooms and into our conscience. How can we ignore the sunken faces of starving children or the carnage brought by war? We cannot and should not. The demand for action to deploy U.S. forces shoulders the American military and political leadership with a moral responsibility that heretofore was not a concern, as our attention was previously focused on the Soviet Union.

Thanks to technological advances in our global transportation system, we can "reach out and touch" more people in more locations in all corners of the world. The more people we reach and the more media attention is devoted to suffering and devastation around the world, the more our transportation system is taxed. Only the United States can respond with the agility and flexibility required to meet these demands.

In his memoirs, Gen. William Tecumseh Sherman wrote, "An army is efficient for action and motion in the inverse proportion to its impedimenta." This maxim is as applicable today as it was during the Civil War. Military forces, whether engaged in a permissive or nonpermissive environment, must minimize their logistical footprint or else slow their "action and motion." The rapid deployment and employment of lighter, more dispersed forces, particularly those well beyond the littorals, place an increased premium on force insertion and timely resupply—resupply from the *air*. In addition, with fewer overseas bases and forward-deployed forces and equipment, timely initial response and sustainment can only be achieved by air mobility.

The United States depends on a flexible and responsive global transportation system that can get American and allied forces to a theater in a timely and decisive manner. Over 2,000 years ago Sun Tzu said, "he who occupies the field of battle first and awaits his enemy is at ease; he who comes later . . . is weary."[3] We cannot afford to be second.

Air mobility allows us to be there first and to control the "battlefield"—it is the air bridge to engagement with aerial refuellers as girders. In the past year alone air mobility forces delivered forces early and decisively to over 30 operations around the world. Air mobility operates 24 hours a day, 365 days a year, in all weather conditions, in all corners of the world, in peace and in war. There are only eleven countries in the world where we did not find American air mobility forces in the past year; two of them did not have airfields! Nothing moves without mobility. Nothing moves quickly and decisively without air mobility.

Notes

1. *Air Mobility Command White Paper* (Washington, D.C.: GPO, 1992), 2.
2. Anthony Cordesman, "The Quadrennial Defense Review and the American Threat to the United States," Manuscript, (Washington, D.C.: CSIS, 1996)
3. Sun Tzu, *The Art of War*, trans. Samuel Griffith (Oxford: Oxford University Press, 1971).

Note: The views presented in this paper are those of the authors as derived from the working group discussions and do not necessarily represent the views of the Department of the Army or the Department of Defense.

7

Technology and the Industrial Base

Ivars Gutmanis

Technological superiority is crucial to air and space dominance in the new millennium, but the key technologies critical to the six core competencies of the United States Air Force cannot be supported by the civilian economy. Among the critical technologies with little civilian support are stealth, counterstealth, space surveillance, survivability of space-based systems, space-based weaponry, targeting, precision drop, short takeoff and landing (STOL), antijam, precision-guided munitions (PGMs), encryption, and information survivability.

Further, reduced defense budgets and the concomitant reductions in both basic research (6.1, 6.2) and the transition programs (6.3a, 6.3b) will inhibit the arrival of new technology and its transition to operational systems. Should the United States rest on its technological laurels, the rest of the world surely will catch up (if not surpass) and cancel out the technological superiority that has permitted U.S. dominance in air and space.

A number of technologies need specific action by the U.S. military to provide necessary improvements. Among these technologies are survivability, air vehicle stealth, mobility assets, space-based survivability, all-weather strike (both targeting and PGMs), and noncooperative Identification Friend or Foe (IFF). Distressingly, these critical technologies by and large are of limited interest to the civilian market.

The technologies identified in this report should receive political and financial support, their export should be undertaken with extreme caution, and a continuous review of both prime and lower-tier suppliers should be made to ensure that a healthy competitive market for such assets remains.

The status of the U.S. industrial base, related trends, and government policies and programs that may affect U.S. air and space power in the new millennium are the focus of this essay. Four issues are of particular relevance to the future of the nation's air and space power:

- The potential contributions of the advanced technologies to each of the six USAF core competencies—i.e., air and space superiority, global attack capability, rapid global mobility, precision engagement ability, information superiority, and agile combat support.
- The technologies critical for enabling the six core competencies.
- The synergistic technologies that may support advances across multiple competencies.
- The characteristics of the evolving industrial base for global defense that may affect the U.S. military's ability to acquire and maintain the required aerospace capabilities.

The Working Group on Technology and the Industrial Base explored three principal areas of analysis that explicitly address these four issues:

- The ability of the consolidating U.S. aerospace industrial base to support alternative modernization programs and yet provide additional resources should unexpected losses occur.
- The potential of selected key technology investment choices to significantly improve the performance of U.S. aerospace forces.
- The implications of Department of Defense initiatives to place reliance on the domestic as well as foreign commercial technologies and systems in order to maintain aerospace superiority, particularly with respect to space-based systems.

Status and Prospects of the U.S. Technology and Industrial Base

The industrial base that supports and advances U.S. air and space power is undergoing substantial changes. These changes and the trends in this base depend mainly on the type and magnitude of defense purchases to maintain and increase U.S. air and space power. U.S. defense expenditures, including those for air and space warfare, will continue to decline unless significant international political developments threaten our defense posture.

Although the nation's technology superiority in air and space activities derives support from the U.S. industrial infrastructure, the industrial base and the technology base are not one and the same.

The technology base includes specific research and development (R&D) activities, both private and governmental, that can be brought to bear in time of need. As a result of the rapid technological evolution in the United States, the supremacy of U.S. air and space power cannot be challenged at this time. It is reasonable to expect that certain air- and space-related technologies will be advanced by appropriate government science and technology (S&T) policies and programs. Our air and space technology advancement, however, will be affected by the reduction in defense-related expenditures. In particular, the DOD proposal to reduce funding of the research, development, testing, and evaluation (RDT&E) budget, which in the decade of the 1980s was about equal to the total engineering and development expenditures of all U.S. civil manufacturing sectors, will retard progress in air- and space-related technology.

Curtailment of research and development in DOD budgets and expenditures will also affect all R&D contract work undertaken by outside entities (e.g., academic institutions, industrial firms, commercial laboratories) for the Department of Defense. That proportion of the R&D effort is considerable. During the past decade, government laboratories contracted out about 60 percent of their applied research and development for procurement programs. The reduction in total DOD R&D expenditures will hurt these contractors. DOD also proposes reductions in the subcategories of Basic Research (6.1), Applied Research (6.2) and Large-scale Experimentation (6.3A). These reductions, unless based on sound analyses, are alarming, as they will hinder scientific advances in air and (especially) space activities related to the nation's defense.

Over the last several years, our R&D and related activities that yield advanced technologies in weapons systems and other defense-related matériel have undergone cardinal changes. At the risk of oversimplification, these changes can be characterized as a major shift from research and development undertaken for the explicit benefit of the nation's defense to R&D conducted for the purpose of advancing the commercial sale of goods and services. The so-called dual-use initiative by DOD has encouraged shifts from "pure" defense-focused R&D to R&D that serve civilian markets. Government initiatives that foster such a shift ought to be balanced by careful consideration of the feasibility of market-driven R&D to meet the nation's defense needs.

The nation's industrial base has experienced changes that replicate changes in research and development. Many key industrial entities that once served the nation's defense effort have been modified to manufacture dual-use goods and render services for civilian markets. The U.S. government is supporting these changes under the dual-use policy. Though recognizing the market forces that encourage, even mandate, such shifts, it is important for the government to carefully consider the need to maintain critical defense-related manufacturing facilities and production lines. This recommendation pertains to prime contractors as well as to lower-tier manufacturing entities.

Air- and Space-Related Technology Issues

The issues concerning the current status and the future prospects of air- and space-related technologies are complex due to the following factors:

- Rapid technological change in several key sectors related to air and space activities.
- DOD emphasis on dual-use goods and services in procurement.
- Emerging international competition in some areas of air and space technology.

Chart 7.1 presents a summary of issues that pertain to the critical technologies from a DOD perspective.

The identification of key air and space technologies and the subsequent advancement of such technologies are of critical importance to the nation's air and space missions in the next millennium. Development and advancement of the identified technologies are of singular importance to the nation's defense posture. The identification of such technologies should govern policies and programs in the selection of primary industrial sectors. It should also play a major role in enacting policies and programs directed toward advancement of the domestic dual-use and commercial industrial activities as well as in foreign procurement.

Chart 7.2 presents a number of technologies critical to the maintenance of the nation's defense posture (listed as USAF core competencies). These technologies were analyzed utilizing a set

Chart 7.1
Issues in Air- and Space-Related Critical Technologies

---

1. The private sector, not DOD, now drives information technology:
   - A similar shift will occur in space and other sectors during the next decade.
   - The good news: off-the-shelf is cheaper, quicker.
   - The bad news: future adversaries will be able to acquire better weapons for little investment.
2. Not all products will be commercially available.
3. Industry is not making up for reduced DOD R&D funding.
4. The trend toward acquisitions and mergers is now moving into lower tiers:
   - This move concentrates capabilities and investment dollars.
   - It may undermine competition or make technologies more competitive internationally.
5. Buying abroad increases the supplier market and may lower costs, but it can undercut domestic vendors, erode the U.S. industrial base, and result in the unexpected denial of critical components because of political and/or economic changes in the source country.

---

of dualities (technical characteristics that may provide merit to a core USAF competency) vis-à-vis functional divisions of the air force's six core competencies, as shown in chart 7.3. Such analyses permit an appraisal of DOD's application of the key technologies needed to leverage leading-edge air power (chart 7.4). The necessity for additional technology advancement is stated in increasingly more critical/demanding requirements for appropriate government policies, programs, and projects as follows:

- Requirement for continuous appraisal.
- Requirement for upgrade.
- Requirement for critical improvement.

The Need to Advance Air- and Space-Related Unique S&T and R&D Activities

The current dual-use goods and technology initiatives of the Department of Defense comprise the nation's policy for the

Chart 7.2
Critical Technologies by USAF Core Competency

Air and Space Superiority

Air

Stealth	All-weather
Counter-stealth	Countermeasures
Improved materials	Avionics
Affordability	Sensors
Precision-strike weapons	Data links
Design allows technology insertion	Automatic target recognition
Radar	Identification Friend or Foe (IFF)
Global Positioning System (GPS)	Defense suppression
Anti-jamming	

Space

Surveillance of and from space	Design allows technology insertion
Communications	On-board power
Survivability	Sensors
Affordability (particularly for launch)	Space-based weaponry
	Launch on demand

Global Attack

Space maneuverability	Targeting, command and control, retargeting
Space-based weaponry	
Positive command and control and recall capability for manned/ unmanned vehicles	Design allows technology insertion
	All-weather
Rapid power projection	Nonlethal

supply of required defense-related matériel during a period of reduced defense expenditures. DOD ought to explicitly consider the reliability of the supply of defense systems that have no demand in the civilian market and thus will not be furnished by dual-use industrial entities. Further, and even more important, advances in certain dual-use technologies, and therefore associated S&T and R&D activities, should not be supported by dual-use firms unless adequate demand for such technology exists in the civilian market. Some of these technologies, however, may represent significant advances in the nation's arsenal.

It is therefore of paramount importance that the government identify such technologies and provide appropriate assistance for their advancement. Chart 7.5 provides a listing of representative technologies and systems.

Chart 7.2 (continued)
Critical Technologies by USAF Core Competency

Rapid Global Mobility

Precision drop
On-the-ground mobility
Asset visibility

Automated and intelligent logistics
Short field takeoff and landing
Air-to-air refueling

Precision Engagement

Anti-jamming
All-weather precision-guided missiles
Deep penetrators
Standoff range
Real-time target identification and post-fire real-time retargeting

Rapid communication to strike force
Battle damage assessment
Survivable munitions
Design allows technology insertion
Unattended ground sensors
Unmanned combat aerial vehicles
Loitering vehicles

Information Superiority

Encryption
Survivability
Anti-jamming
Sensors
Integration
Design allows technology insertion
Protection against self-jamming
Decision aids
Dissemination of information to tactical level
Information filters

Compression technology for systems
Secure data link
Consequence management of the interaction of the commercial and defense systems
Information warfare (jamming, sending false information, accessing enemy information)
Redundancy
Interoperability

Agile Combat Support

Automated and intelligent logistics
Minimization of in-country inventory

Asset viability

Chart 7.3
Critical Technology Dualities

manned	*versus*	unmanned
air	*versus*	space
shorter-range	*versus*	longer-range
reusable	*versus*	expendable
brilliant	*versus*	just smart enough
virtual	*versus*	physical
survivability	*versus*	vulnerability

Chart 7.4
Requirements for DOD Action on Selected Air- and Space-Related Technologies by Competency Area

Core Competency	DOD Action Required:		
	Continuous appraisal	*Upgrade*	*Critical improvement*
Air Superiority and Global Attack	• Unmanned capabilities • Global Positioning Systems (GPSs)	• Helmet-mounted systems • Ejection seats for smaller pilots • All-weather systems • Search, recognition, and attack functions for weapons systems • Counter-stealth • Infrared signatures • GPSs	• Survivability • Cruise missile defense
Space Superiority	• Launch vehicle production lines • Vulnerability and reliability	• Space-based systems • Space-surveillance capabilities • Launch on demand	• Survivability
Precision Engagement	• Counter-counter measures	• Cost containment	• All-weather capability • Hard-target penetration • Featureless targeting

Chart 7.4 (continued)
Requirements for DOD Action on Selected Air- and Space-Related Technologies by Competency Area

Core Competency	DOD Action Required:		
	Continuous appraisal	*Upgrade*	*Critical improvement*
Rapid Global Mobility	• Supply system management in airlift • Tagging of supplies	• Autonomous mobile response capability	• Precision air drop • Survivability
Agile Combat Support	• Tactical supply capability • Precision delivery to remote units	• Systems monitoring	• In-theater improvement capability for low-volume production
Information Superiority	• Persistent combat management of variated systems	• Adequate data links and spectrum management	• C^3 (command, control, and communication) of multiple UAVs • Bio-sensor • IFF • Control and dissemination

Chart 7.5
Air- and Space-Related Critical Technologies with Limited Interest to the Civilian Marketplace

1. Stealth-related technologies
2. Technologies focused on the electronic hardening of microelectronic components
3. Ballistic protection technologies
4. Explosives and ammunition
5. Advanced detectors
6. Anti-chemical and anti-biological warfare agents

Air- and Space-Related Industrial Base Issues

Major changes in the air- and space-related industrial base have occurred during the last decade and additional changes are anticipated in the near future. A summary of the issues and requirements resulting from these changes is presented in chart 7.6.

The principal determinants of past and current changes in the nation's air and space industrial base are, and will continue to be, a reduction in defense expenditures. Differing defense budget forecasts have been developed but all indicate a substantial decline in the air- and space-related defense budget categories. Assuming, then, continued budget reductions one must also assume their continued impact on the business of prime as well as lower-tier contractors that support the nation's air and space programs. The government should continuously review the status (availability, capacity, and technology—both process and product) of the air- and space-related industrial base by industry sector (and corporate entity) corresponding to the critical technologies for the U.S. Air Force's six core competencies.

The current trends affecting the defense industrial base also merit continuous review of both the prime and lower-tier industrial firms that manufacture critical parts for the air and space systems. For optimum results, the Department of Defense should develop a comprehensive program for the supply of critical parts by both prime and lower-tiers contractors. In addition,

Chart 7.6
Industry's Role in Air and Space Technology:
Issues and Requirements

Industry is an increasingly critical participant in both the military and business revolutions. The following are five general prescriptions for maintaining a responsive and robust industrial base.

1. Develop technologies to provide a twenty-first century advantage.
2. Embrace concepts for new systems:
   - affordable manufacturing and logistics
   - reduced cycle times
   - rapid technology infusion.
3. Reduce costs through dual-use, commercial off-the-shelf (COTS) technologies and civil-military integration.
4. Create and maintain a higher-quality, more responsive, lower-cost infrastructure:
   - transition from a supporting role to a leadership role
   - maintain profits/selected skill base despite diminishing orders.
5. Support DOD efforts to take a global view and increase reliance on commercial capabilities and practices.

U.S. facilities and industrial manufacturing processes and production lines should be maintained in a "warm" state, even if no actual manufacturing or production activities take place in their facilities.

Procurement Issues

The acquisition reform enacted by the Department of Defense during 1994 and 1995 has significantly changed the operations of defense procurement. These changes, combined with the dual-use initiative, emphasize acquisition of government off-the-shelf (GOTS), non-developmental item (NDI) procurement, and commercial off-the-shelf (COTS) manufactured in accordance with commercial standards rather than under military specification (MILSPEC) regulations. The principal goal of these procurement changes is to ensure greater affordability for defense-related goods and services. The principles and regulations of the acquisition reform should accommodate the rapidly diminishing defense-related industrial base in general and supply the required materiél for the nation's air and space activities in particular. Such accommodation is possible only if the acquisition regulations explicitly allow, even encourage, certain exemptions from normal government procurement procedures when dictated by the characteristics of the changing defense industrial base.

Especially in those cases where the normal defense, dual-use, or market forces fail to operate, the following two exemptions to the prescribed procurement procedures should be considered:

- Sole-Source Supplier(s): Acquisitions needs occasionally may justify, in terms of cost and efficiency, sole-source procurement. Such procurement should be evaluated and approved when mandated by cost and efficiency considerations.

- Preferred Suppliers: The Department of Defense should consider the benefits of preferred suppliers for appropriate defense-related materiél. The utilization of preferred suppliers (as opposed to open competition) should be approved when mandated by cost, efficiency, time lines, and secure supply factors.

Finally, a more "open" relationship between the DOD procurement agencies and the supplier base is desirable. The efficiency and cost of DOD procurement activities are significantly hindered by the unwillingness of DOD procurement agencies to discuss DOD requirements with potential suppliers.

Impact of Industry Consolidation on the Air and Space Activities

During the last five or so years, the U.S. defense-related industrial base has undergone significant consolidation, principally resulting from reductions in the national defense budget.

These reductions in defense spending are taking place during a period of potentially enormous technological advances. But the conversion of these potential advances into fielded air and space weapons systems requires a robust domestic industrial base.

Beginning in 1990, an increasing number of large U.S. defense-related firms have initiated mergers, acquisitions, or other types of business collaboration in order to enter commercial markets or to continue producing weapons systems and other defense materiél.

The importance of mergers as a strategy for survival has grown, as indicated by the reported estimated value of the defense firms undertaking them. In 1992 their value was estimated at $300 million; in 1992, $740 million; in 1993, $6.4 billion; in 1994, $14.2 billion; and in 1995, down to $5.5 billion, only to increase to an estimated $14 billion in 1996. The industrial entities created by these mergers represent very large, vertically integrated defense firms capable of producing a wide variety of defense-related products.

To date, most of the mergers in the defense industry have been successful, but this approach may not represent the optimum long-range solution for maintaining a viable air- and space-related industrial base. Mergers may reduce competition and industrial base depth and in the long run may significantly increase the cost of defense materiél. Additionally, vertical integration of some firms will adversely impact the lower-tier entities. To alleviate the harmful effects to smaller firms of such restructuring, the government should enact policies and programs that assist lower-tier firms in their conversion to civilian markets or dual production.

The Use of Foreign Sources of Supply

The dual-use initiatives by the Department of Defense also suggest dual-use product procurement from foreign sources. Such procurement may reduce the cost of acquisition or may be required because of the unique availability of specialized foreign products. These transactions strengthen bonding between allies, improve interoperability, and enhance foreign military sales.

The United States, however, should be cautioned against significant reliance on foreign sources for critical goods and services. Arguably, the international trade agreements among about sixty-seven nations (including the United States), which are administrated by the World Trade Organization (WTO), have reduced trade conflicts and risks associated with the procurement of foreign goods and services for the nation's defense needs. The multitude of coproduction and offset programs maintained by the Department of Defense may also minimize risks of procurement from foreign sources. Conversely, there is ever increasing evidence that the nations that comprise the

European Union (EU), as well as others (notably Japan, Korea, Taiwan, and Brazil), are engaged in building a "native" R&D and manufacturing capability for their own use and, increasingly, for international sales. It has also been estimated that out of the total annual worldwide defense-related sales of some $18 billion, European nations account for approximately $6 billion and are systematically stepping up their defense-related sales to other nations.

Increased competition should also result from a score of transnational mergers, purchases, and joint undertakings of defense-related programs in Europe. Some of the participants in this wave of reconstruction represent the principal European air and space entities and are potential competitors to U.S. firms. These international air and space matériel ventures in Europe may in the future compete directly with U.S. firms in sales to Europe, the North Atlantic Treaty Organization (NATO), and Third-World countries. Another concern is that air- and space-related technology transfer from the United States without appropriate policy protection may provide an advantage to these potential competitors and may harm U.S. contractors.

Several major obstacles to DOD procurement overseas remain. The most important of these are extraterritoriality; intellectual property including patent-related issues; and intra-EU weapons projects. Among these three possible barriers to foreign procurement and sales of air and space goods (parts, components and systems), the extraterritoriality provisions and intra-EU weapons projects are the most important. In the case of extraterritoriality, we should continue to preclude international trade in air and space systems that may ultimately be acquired by potential adversaries.

With regard to their own weapons development, it is possible that foreign countries, including NATO members, may refuse to sell certain defense-related goods developed for their own weapon systems. The United States, in turn, should continue to halt the sale of certain air and space warfare-related components to the other nations. We should conclude appropriate agreements with other countries to preclude mutually harmful developments.

Conclusion

It is apparent that defense expenditures, and thus the aerospace industrial base, will decline unless (1) a significant military or economic (e.g., denial of oil) threat materializes, or (2) futuristic-thinking senior U.S. policy and budget makers recognize major potential threats and insist on preparing to meet them. Subsequently, given sufficient time and money, there is little doubt that the consolidated U.S. aerospace industrial base can, as it has before, "rise to the occasion." However, it is unlikely that a serious aggressor would permit us the luxury of "time." Therefore, we must maintain a ready, trained, and equipped military establishment backed by a viable and adaptable industrial base. To quote Thomas Jefferson, "Eternal vigilance is the price of liberty."

Because the U.S. capability to support alternative modernization programs cannot be assured by relying on industrial capabilities driven by commercial interests, the U.S. government must continue to have the cutting edge of defense technology by funding R&D and modernization programs to ensure that suppliers at all tiers do not gravitate to only those areas that promise commercial profit. Where COTS items meet military requirements, they should be phased in over time. Where maintaining a MILSPEC is less costly than conversion to COTS, the MILSPEC should prevail.

Although many technology areas are identified for special consideration, the ones that will contribute most to maintaining air and space superiority and that will not be accomplished by relying on pure commercial interests are those related to stealth and deception (especially invisibility to sensors other than just radar); hardening of microelectronic components in all platforms; lightweight shielding (i.e., armor protection) of personnel and equipment; development of nonjammable, multi-spectrum kill-vehicle sensors delivering capable, target-specific (e.g., deeply buried command centers) munitions; and neutralization of chemical/biological agents. Although there are valid reasons to consider foreign sources of supply, these suppliers should be used only to satisfy specific U.S. policy and/or foreign marketing objectives while not sacrificing the U.S. industrial base's capability to rapidly fill the gap should the foreign source suddenly be denied.

Abbreviations and Acronyms

ABIS	Advanced Battleship Information System
ABL	airborne laser
ABM	antiballistic missile
ACTD	advanced concept and technology demonstration
AEF	Air Expeditionary Force
AFIWC	Air Force Information Warfare Center
AFRES	Air Force Reserve
ANG	Air National Guard
ATACMS	Army Tactical Missile System
ATO	air tasking order
AWACS	airborne warning and control system
AWE	Advanced Warfare Experiment
BDA	bomb damage assessment
C^3	command, control, and communications
C^4	command, control, communications, and computers
C^4ISR	command, control, communications, computers, intelligence, surveillance, and reconnaissance
CAP	combat air patrol
CEC	cooperative engagement capability
CEP	circular error probable
CINC	commander in chief
CONUS	continental United States
COTS	commercial off-the-shelf
DARO	Defense Airborne Reconnaissance Office

DAWMS	Deep Attack Weapons Mix Study
DOD	Department of Defense
DR&E	Defense Research and Engineering
DSP	Defense Support Program
ECM	electronic countermeasure
EMP	electromagnetic pulse
EU	European Union
EW	electronic warfare
FEBA	forward edge of the battlefield
GATS/GAM	GPS-assisted targeting system and GPS-assisted mission planning
GBS	global broadcast system
GCCS	global command and control system
GII	global information infrastructure
GOTS	government off-the-shelf
GPS	global positioning system
GWAPS	Gulf War Airpower Survey
HALO/HALE	high-altitude, low-observable and high-altitude, long-endurance
IADS	integrated air defense system
ICBM	intercontinental ballistic missile
IFF	Identification Friend or Foe
INS	inertial navigation system
IO	information operations
IS	information superiority
ISR	intelligence, surveillance, and reconnaissance
IT	information technology
IW	information warfare
JCS	Joint Chiefs of Staff
JDAM	joint direct attack munition
JFACC	joint forces air component commander
JSTARS	Joint Surveillance Target Attack Radar System
JTIDS	Joint Tactical Information Distribution System

LANTIRN	Low-Altitude Navigation and Targeting Infra-red Night
LGB	laser-guided bomb
MILSPEC	military specification
MOE	measure of effectiveness
MRS BUR	Mobility Requirements Task Force Bottom-Up Review
NATO	North Atlantic Treaty Organization
NBC	nuclear, biological, and chemical
NCA	National Command Authority
NCO	noncommissioned officer
NDI	non-developmental item
NDP	National Defense Panel
NII	national information infrastructure
OOTW	operations-other-than-war
PERSTEMPO	personnel tempo
PGM	precision-guided munition
PGW	precision-guided weapon
PREPO	pre-positioned
QDR	Quadrennial Defense Review
R&D	research and development
RDT&E	research, development, testing, and evaluation
RMA	Revolution in Military Affairs
S&R	surveillance and reconnaissance
S&T	science and technology
S/APOD	sea/aerial ports of debarkation
SAC	Strategic Air Command
SAM	surface-to-air missile
SBIRS	space-based infrared system
SDI	Strategic Defense Initiative
SDIO	Strategic Defense Initiative Office
SEAD	suppression of enemy air defense
STOL	short takeoff and landing
TBM	theater ballistic missile

TFR	terrain-following radar
TMD	theater missile defense
TRI-TAC	Tri-Service Tactical Communications Program
UAV	unmanned aerial vehicle
UCAV	unmanned combat aerial vehicle
USA	U.S. Army
USAF	U. S. Air Force
USMC	U.S. Marine Corps
USN	U.S. Navy
USTRANSCOM	United States Transportation Command
WMD	weapons of mass destruction
WTO	World Trade Organization
WWMCCS	World-Wide Military Command and Control System

Working Groups

Working Group on Strategy
Chairmen:

General Michael Dugan
United States Air Force (Ret.)

Admiral Ronald "Zap"
  Zlatoper, USN (Ret.)
Sanchez Computer Associates

Members:

Andy Andreson
Raytheon

Freda Bryce
Global Technologies Incorporated

John Conoway
Air National Guard (Ret.)

Jeffrey Cooper
SAIC

Daniel Gouré
CSIS

Charles Herzfeld
CSIS

John Hillen
Council on Foreign Relations

Bruce Jackson
Lockheed Martin

Bruce Klassen
A.T. Kearney

Dewey Mauldin
CSIS-USMC

Ervin Rokke
National Defense University

Don Snider
United States Military Academy

Dov Zakheim
System Planning Corporation

Working Group on Air and Space Superiority
Chairman:

Dr. Richard Hallion
United States Air Force Historian

Members:

Jimmie Adams
Lockheed Martin

Mason Botts
Raytheon

Bron Burke
McDonnell Douglas

Tom Burke
TRW

Richard Dawe
USN-CSIS

Joe Draham
EDS

Michael Irish
Simulation Technologies Inc.

Jeffry Jackson
USAF-CSIS

Shephard Hill
Boeing

Gordon Middleton
TASC

Benjamin Lambeth
RAND

Michael Loh
United States Air Force (Ret.)

David McCurdy

Al Pruden
Lockheed Martin

Don Rakestraw
VII Inc.

Leighton Smith
Center for Naval Analyses

Bill Switzer
TRW

Working Group on Global Attack and Precision Strike

Chairman:

Arnold Punaro
SAIC

Members:

T. I. Anderson
Boeing

Stephen Cambone
CSIS

Roger Collins
Lockheed Martin

Bill Dalecky
McDonnell Douglas

Don Fredricksen
Hicks & Associates

Patrick Garrity
Los Alamos National Laboratory

Roger P. Heinisch
Alliant Techsystems

Jeffry Jackson
USAF-CSIS

Mike Kerby
Raytheon

George Plummer
United Defense LP

Steve Roemerman
Texas Instruments

Earl Rubright
CENTCOM

Eric Taylor
Raytheon

Working Group on Information Superiority
Chairman:
Vice Admiral Mike McConnell, USN (Ret.)
Booz-Allen & Hamilton

Members:

John Alger
Kaman Sciences

Kenneth Allard
Cyber Strategies

Jana Cira
Hughes

Charles Culbertson
ISX

Grady Culbertson
TRW

Chuck Cunningham
Defense Intelligence Agency

Ryan Henry
CSIS

Carl O'Berry
Motorola-USAF (Ret.)

Edward Peartree
CSIS

Patrick Reidy
Newport News Shipbuilding

Marijean Seelbach
TRW

Edward Shirley
USAF (Ret.)

Gilbert Siegert
Space Ventures Consulting

David Signori
DARPA

Rich Spooner
Lockheed Martin

Peter Wilson
RAND

Working Group on Mobility and Support
Chairmen:

General H.T. Johnson
United States Air Force (Ret.)

Brigadier General Tom Mikolajcik
United States Air Force (Ret.)

Members:

Anthony Burshnick
United States Air Force (Ret.)

Richard Dawe
USN-CSIS

Robert Ensslin
United States Army National Guard (Ret.)

Robert Ewart
McDonnell Douglas

Calvin Franklin
Engineering Systems Consultants

Harry Griffith

Alfred Hansen
Lockheed Martin

Keith Hutcheson
LEH Consultants

Robert McClure
USA-CSIS

William Tuttle
Logistics Management Institute

John Pestonjee
Texas Instruments

Harlan Ullman
Center for Naval Analyses

George Sinks
Logistics Management Institute

Jessica Wright
USANG-CSIS

Working Group on Technology and the Industrial Base
Chairman:

Hal Howes
Howes & Associates, Ltd.

Members:

Alan Bernard
MIT Lincoln Laboratory

James Lindenfelser
TASC

Robert Bott
McDonnell Douglas

Dewey Mauldin
USMC-CSIS

Cal Coolidge
Texas Instruments

Charles McGrail
MVM Associates

Ivars Gutmanis
Hobe Corporation

Austin O'Toole
McDonnell Douglas

John Jaquish
Jaquish Associates

Rich Palaschak
Alliant Techsystems

About the Contributors

C. Kenneth Allard, a former U.S. Army colonel, is chief executive officer of Cyber Strategies Inc., a technology consulting firm specializing in information warfare. His military career included service on the West Point faculty, as special assistant to the army chief of staff, and as dean of students at the National War College. He also helped craft two major Pentagon reforms, the Goldwater-Nichols Act of 1986 and the Federal Acquisition Streamlining Act of 1994. A command and control expert, he served in 1996 on special assignment in Bosnia with the U.S. 1st Armored Division. Dr. Allard's publications include *Command, Control and the Common Defense*, winner of the 1991 National Security Book Award. He holds a Ph.D. from the Fletcher School of Law and Diplomacy and an MPA from Harvard University and is an adjunct professor in the National Security Studies Program, Georgetown University.

Stephen A. Cambone joined CSIS in June 1993 following three years in the Office of the Secretary of Defense as director of the strategic defense policy office. He began his career at Los Alamos National Laboratory, where he specialized in strategic and theater nuclear weapons requirements and related arms control issues. He moved to the private sector as a defense analyst with SRS Technologies. As director of strategic defense policy, he was a major contributor to President George Bush's decision to refocus the Strategic Defense Initiative program in 1991 and developed the concept for a global protection system. He was also a member of the group appointed by the president to discuss the global protection system with Russia, U.S. allies, and other states. Dr. Cambone holds M.A. and Ph.D. degrees from Claremont Graduate School, Claremont, California, and a B.A. from the Catholic University of America.

About the Contributors

Saxby Chambliss was elected to his second term in the House of Representatives in November 1996 as representative of Georgia's 8th District. He has been a member of both the National Security and Agriculture committees of the House and is cochairman of the Congressional Air Power Caucus. He was also one of two sophomore representatives selected to serve on the Republican Steering Committee, which is responsible for delegating all committee assignments. Before undertaking a political career, Mr. Chambliss was a businessman and attorney specializing in business and agricultural law. In 1990 he was appointed to the Disciplinary Review Panel of the Georgia state bar. He received a bachelor's degree in business administration from the University of Georgia and earned his J.D. from the University of Tennessee College of Law.

Jeffrey R. Cooper is director of the Center for Information Strategy and Policy at Science Applications International Corporation (SAIC). His 30-year career in defense analysis has included positions at the Hudson Institute, SRS Technologies, the Johns Hopkins University School of Advanced International Studies, the U.S. Arms Control and Disarmament Agency, and the Department of Energy as assistant to Secretary of Energy James Schlesinger. Mr. Cooper's recent work has focused on information warfare and the revolution in military affairs. His writings include *Another View of the Revolution in Military Affairs* (1994) for the U.S. Army War College. He received his undergraduate and graduate education at the Johns Hopkins University.

Ronald R. Fogleman recently retired as chief of staff of the United States Air Force. As chief, General Fogleman served as the senior uniformed U.S. Air Force officer responsible for the organization, training, and equipage of 750,000 active duty, Guard, Reserve, and civilian forces serving in the United States and overseas. As a member of the Joint Chiefs of Staff, he and the other service chiefs functioned as military advisers to the secretary of defense, the National Security Council, and the president. A command pilot and a parachutist, General Fogleman flew 315 combat missions and has amassed more than 6,500 flying hours in fighter, transport, tanker, and rotary wing aircraft. A 1963 graduate of the U.S. Air Force Academy, he holds a master's degree in military history and political science from Duke University.

About the Contributors   169

Daniel Gouré joined CSIS in 1993 after two years in the Office of the Secretary of Defense as director of the Office of Strategic Competitiveness. He began his career with the U.S. Arms Control and Disarmament Agency, specializing in regional arms control issues. He then moved to the private sector as a defense analyst with such firms as Systems Planning Corporation, R&D Associates, Science Application International Corporation, and SRS Technologies. In 1985, while a member of the staff of the Center for Naval Analyses, he was director of the threat panel for the Strategic Defense Initiative Organization-sponsored pilot architecture study. Dr. Gouré holds M.A. and Ph.D. degrees from the Johns Hopkins University. He also has a B.A. from Pomona College. He has been a lecturer at the National War College, Air War College, Naval War College, and the Naval Post Graduate School.

Ivars Gutmanis is chief operating officer of the Hobe Corporation, a management consulting firm established in 1973 and located in Washington, D.C. Prior to that he was director of the resource analysis division at the National Planning Association from 1960 to 1973. Dr. Gutmanis has been a consultant to the United Nations, the OECD, the National Academy of Sciences, Resources for the Future, the Brookings Institution, and other organizations on issues related to technology, manufacturing, the defense industrial base, and national security. He received his Ph.D., with a focus on technology development among OECD member countries, at the University of Durham, United Kingdom.

Richard P. Hallion has been the Air Force historian at Headquarters U.S. Air Force since 1991, where he is responsible for directing the worldwide U.S. Air Force historical and museum programs. Dr. Hallion has broad experience in museum development, historical research, and management analysis and has served as a consultant to a variety of professional organizations. He is the author of 14 books relating to aerospace history, and he teaches and lectures widely.

Keith A. Hutcheson is president of LEH Consultants, Inc., a Fairfax, Virginia-based company focused on strategic military and humanitarian assistance matters. He is a retired U.S. Air Force pilot and combat veteran of Operations Desert Storm and

Just Cause. Before his retirement he served two Pentagon assignments in Air Force legislative liaison activities and one in the strategy division, the so-called Skunkworks. During his tenure in the Skunkworks, Mr. Hutcheson wrote the 1996 white paper "Air Force Strategy and Force Structure." He holds a M.S.S.M. from the University of Southern California and a M.M.S. from the Marine Corps University.

Michael Irish is director of strategic business development for Simulation Technologies Incorporated (STI), an information technology company that specializes in modeling, simulation, software, and war-gaming. Mr. Irish is responsible for STI's new-concept development, long-range strategy, and strategic alliances projects. Before joining STI, he was a research associate at the Economic Strategy Institute, where his work focused on information technology, public- and private-sector partnerships, and research and development in critical technology areas. He is coauthor and editor of several publications, including *Shared Investment-Shared Return: Industry/Government Technology Programs*. Mr. Irish served in the U.S. Marine Corps Reserve and has a degree in international relations from San Francisco State University.

Jeffry A. Jackson has served as a lieutenant colonel in the U.S. Air Force for 17 years and was the USAF military fellow at CSIS in 1996–1997. An F-16 fighter pilot, Lt. Colonel Jackson served as the operations officer for the 56th Training Squadron at Luke Air Force Base, Arizona. His previous flight assignments include tours at Hill Air Force Base, Utah, and Kunsan Air Base, Republic of Korea, where he was F-16 flight commander, assistant operations officer, and instructor pilot. From 1991 to 1994, Lt. Colonel Jackson commanded a cadet squadron and group and taught international relations and U.S. national security policy at the U.S. Air Force Academy. A Rhodes Scholar at Oxford University in 1981, Jackson completed his philosophy, politics, and economics degree there. He also holds a B.S. in electrical engineering from the Air Force Academy, where he also majored in the humanities.

Robert McClure is a lieutenant colonel in the U.S. Army and served as the U.S. Army military fellow to CSIS in 1996–1997. Lt. Colonel McClure came to CSIS after he commanded a construction engineer battalion at Ft. Steward, Georgia, and served as commander of a combined Canadian-American engineer battalion that built base camps for UN forces in Haiti in 1995. His previous assignments include more than 10 years with units throughout Germany and teaching international relations as an assistant professor in the department of social sciences at West Point. A 1976 graduate of the U.S. Military Academy, Lt. Colonel McClure holds masters degrees in public administration from Harvard and in systems management from the University of Southern California and has studied as an Olmsted Scholar at Gutenberg University in Germany.

Carl O'Berry is vice president and director of planning and information technology at Motorola, where he is responsible for information technology architectures and roadmaps, new business development, and leadership of innovation and process reengineering. Previously he was deputy chief of staff for command, control, communications, and computers (C^4) for the U.S. Air Force and commander of the Rome Air Development Center, the largest research and development center for the U.S. Air Force. He received his B.S. in electrical engineering from New Mexico State University and his M.S. in systems management from the Air Force Institute of Technology.

C. Edward Peartree is an analyst with the Office of Strategic Policy and Negotiations, U.S. Department of State. Until recently, he was a research associate in Political-Military Studies at CSIS, where he worked on a multiyear study of information warfare, leading-edge warfare, and the impact of the information revolution on international and national security. He is coeditor of *The Information Revolution and International Security* (forthcoming, CSIS). He holds a B.A. from the Johns Hopkins University and an M.A. from the George Washington University Elliott School of International Affairs.

Christopher M. Szara is a research associate in the Political-Military Studies Program at CSIS, where he has worked on defense science and technology, undersea warfare, information warfare, and various issues relating to the future of aerospace power. He received his B.S./B.A. in computer engineering from Bucknell University and his M.A. in national security studies from Georgetown University.

Charles D. Vollmer is the president of VII Inc., which specializes in doctrine, strategic planning, C^4ISR, and synthetic environments for military clients, as well as distance learning and information security. As a former partner at Booz-Allen & Hamilton Inc., he led one of the four largest U.S. consortia privatizing industry in the former Soviet Union and cofounded the Community Learning & Information Network. As vice president of General Dynamics, he founded the Defense Initiative Organization, which specializes in emerging national initiatives. At McDonnell Douglas he was on the initial design teams for the F-15E and Stealth. In service with the U.S. Air Force and Air National Guard, he accumulated more than 3,000 hours in fighters and flew 175 combat missions in Southeast Asia. He holds degrees from the U.S. Air Force Academy and Northern Arizona University and attended MIT's Sloan School for Senior Executives.